Glencoe McGraw-Hill

Math Connects

Course 2

Noteables™

Interactive Study Notebook with Foldables®

Contributing Author
Dinah Zike

Consultant
Douglas Fisher, Ph.D.
Professor of Language and Literacy Education
San Diego State University
San Diego, CA

The McGraw-Hill Companies

Send all inquiries to:
The McGraw-Hill Companies
8787 Orion Place
Columbus, OH 43240-4027

Printed in the United States of America.

ISBN: 978-0-07-890237-6
MHID: 0-07-890237-1

Math Connects: Concepts, Skills, and Problem Solving, Course 2
Noteables™: Interactive Study Notebook with Foldables®

3 4 5 6 7 8 9 10 009 17 16 15 14 13 12 11 10 09

Contents

Contents

Organizing Your Foldables

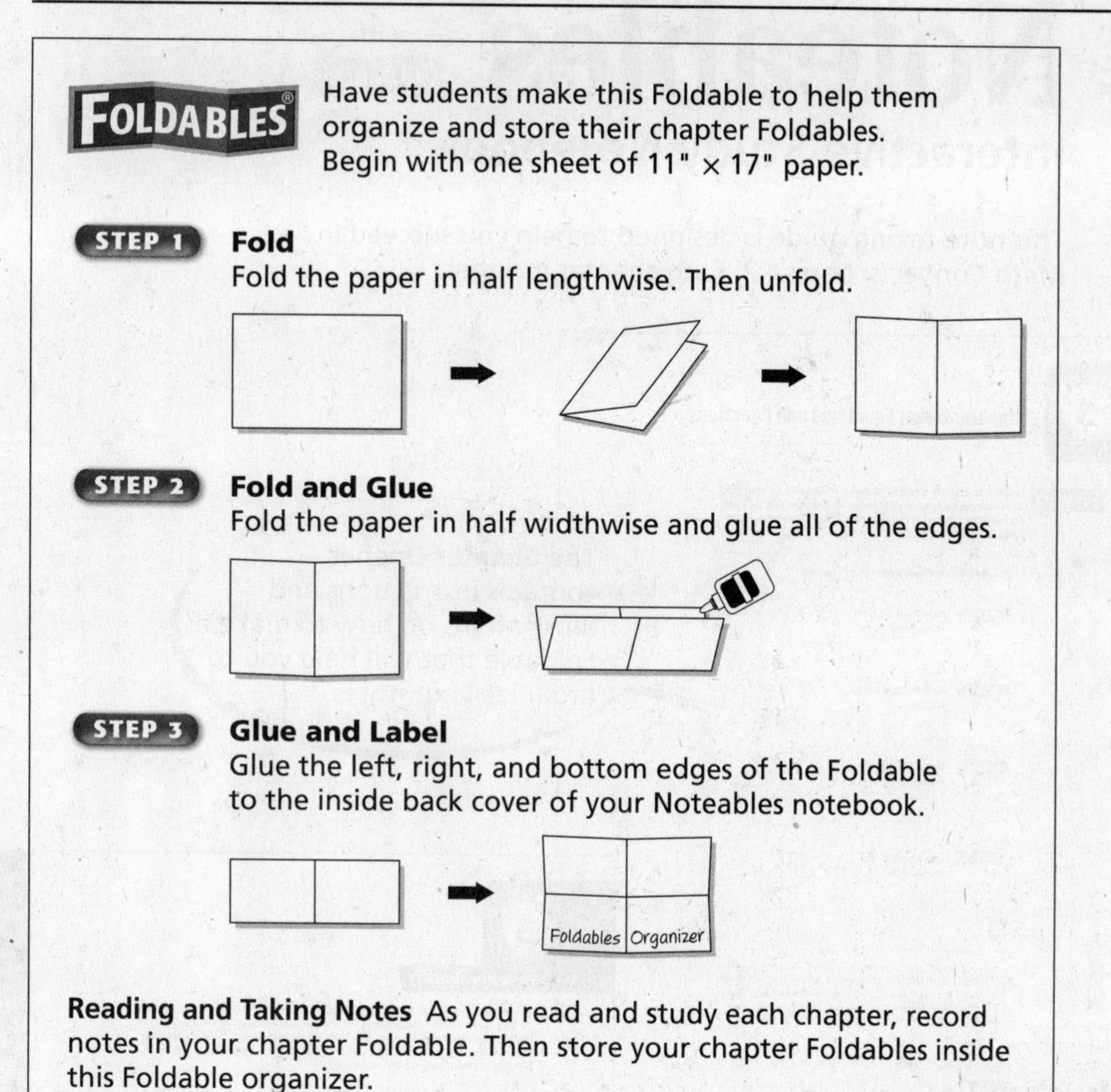

Have students make this Foldable to help them organize and store their chapter Foldables. Begin with one sheet of 11" × 17" paper.

STEP 1 **Fold**
Fold the paper in half lengthwise. Then unfold.

STEP 2 **Fold and Glue**
Fold the paper in half widthwise and glue all of the edges.

STEP 3 **Glue and Label**
Glue the left, right, and bottom edges of the Foldable to the inside back cover of your Noteables notebook.

Reading and Taking Notes As you read and study each chapter, record notes in your chapter Foldable. Then store your chapter Foldables inside this Foldable organizer.

Using Your Noteables™ Interactive Study Notebook

This note-taking guide is designed to help you succeed in *Math Connects*, Course 2. Each chapter includes:

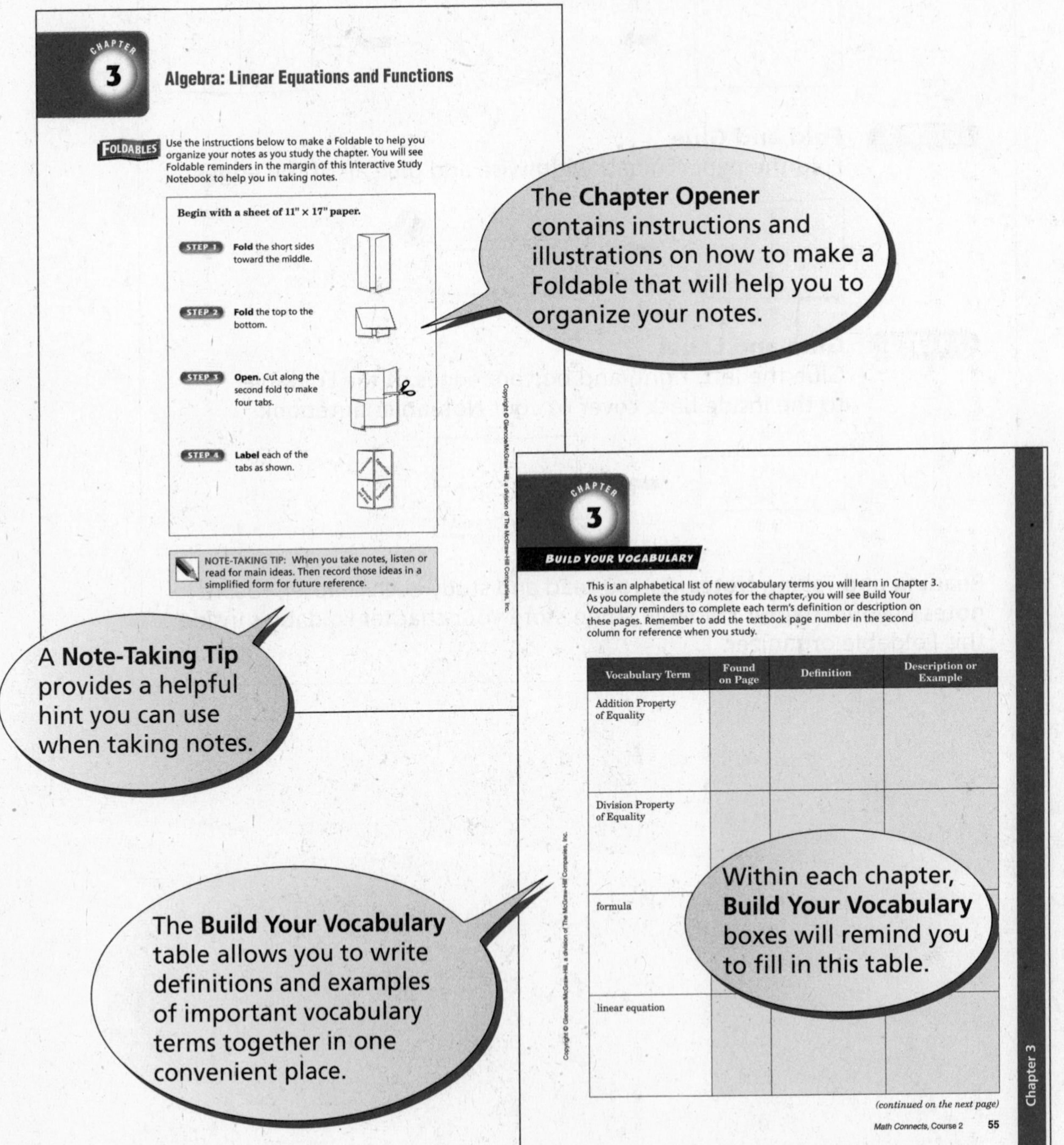

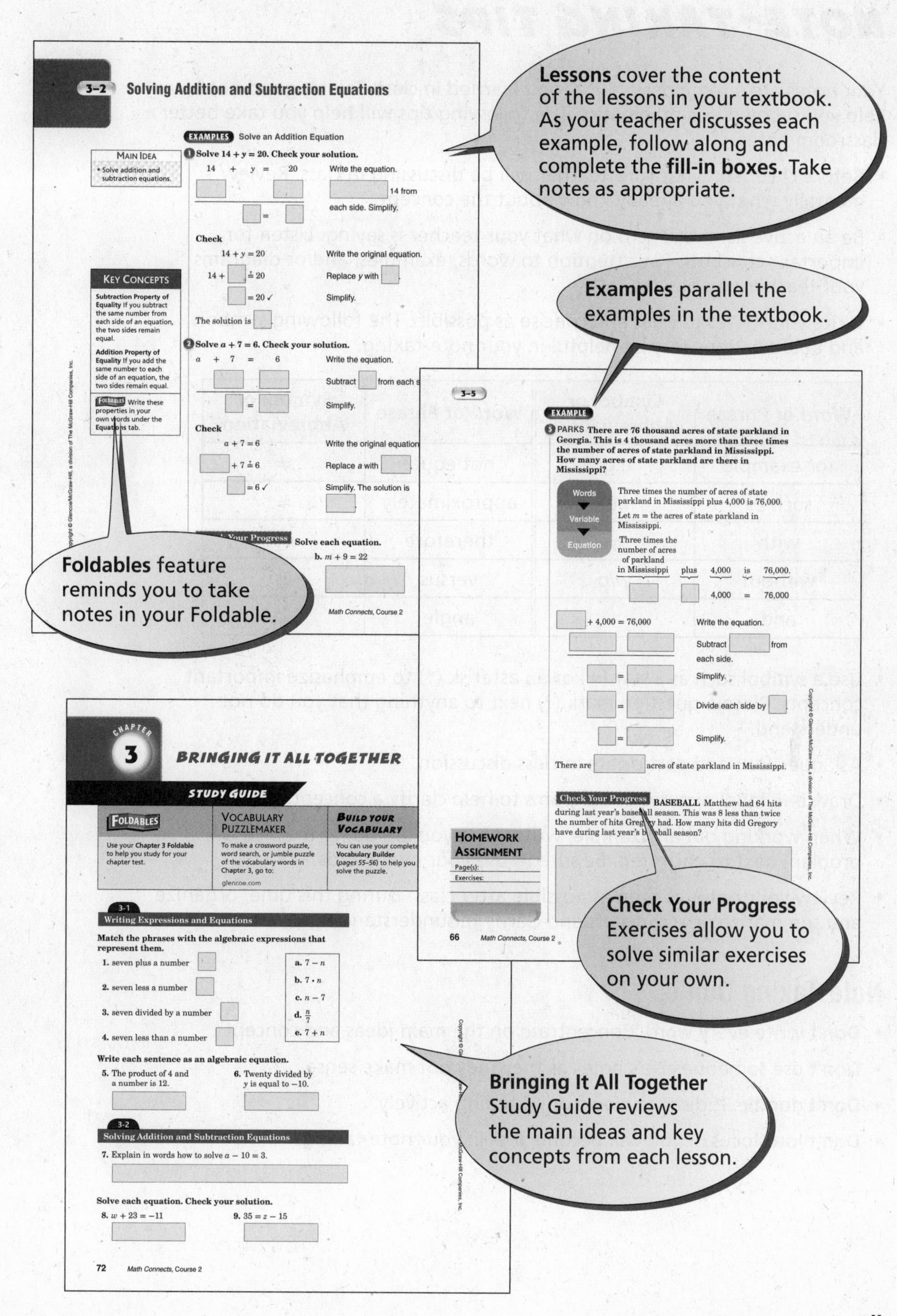
Lessons cover the content of the lessons in your textbook. As your teacher discusses each example, follow along and complete the fill-in boxes. Take notes as appropriate.
Examples parallel the examples in the textbook.
Foldables feature reminds you to take notes in your Foldable.
Check Your Progress Exercises allow you to solve similar exercises on your own.
Bringing It All Together Study Guide reviews the main ideas and key concepts from each lesson.

NOTE-TAKING TIPS

Your notes are a reminder of what you learned in class. Taking good notes can help you succeed in mathematics. The following tips will help you take better classroom notes.

- Before class, ask what your teacher will be discussing in class. Review mentally what you already know about the concept.
- Be an active listener. Focus on what your teacher is saying. Listen for important concepts. Pay attention to words, examples, and/or diagrams your teacher emphasizes.
- Write your notes as clear and concise as possible. The following symbols and abbreviations may be helpful in your note-taking.

Word or Phrase	Symbol or Abbreviation	Word or Phrase	Symbol or Abbreviation
for example	e.g.	not equal	$\neq$
such as	i.e.	approximately	$\approx$
with	w/	therefore	$\therefore$
without	w/o	versus	vs
and	+	angle	$\angle$

- Use a symbol such as a star (★) or an asterisk (*) to emphasize important concepts. Place a question mark (?) next to anything that you do not understand.
- Ask questions and participate in class discussion.
- Draw and label pictures or diagrams to help clarify a concept.
- When working out an example, write what you are doing to solve the problem next to each step. Be sure to use your own words.
- Review your notes as soon as possible after class. During this time, organize and summarize new concepts and clarify misunderstandings.

Note-Taking Don'ts

- **Don't** write every word. Concentrate on the main ideas and concepts.
- **Don't** use someone else's notes as they may not make sense.
- **Don't** doodle. It distracts you from listening actively.
- **Don't** lose focus or you will become lost in your note-taking.

Introduction to Algebra and Functions

Use the instructions below to make a Foldable to help you organize your notes as you study the chapter. You will see Foldable reminders in the margin this Interactive Study Notebook to help you in taking notes.

Begin with eleven sheets of notebook paper.

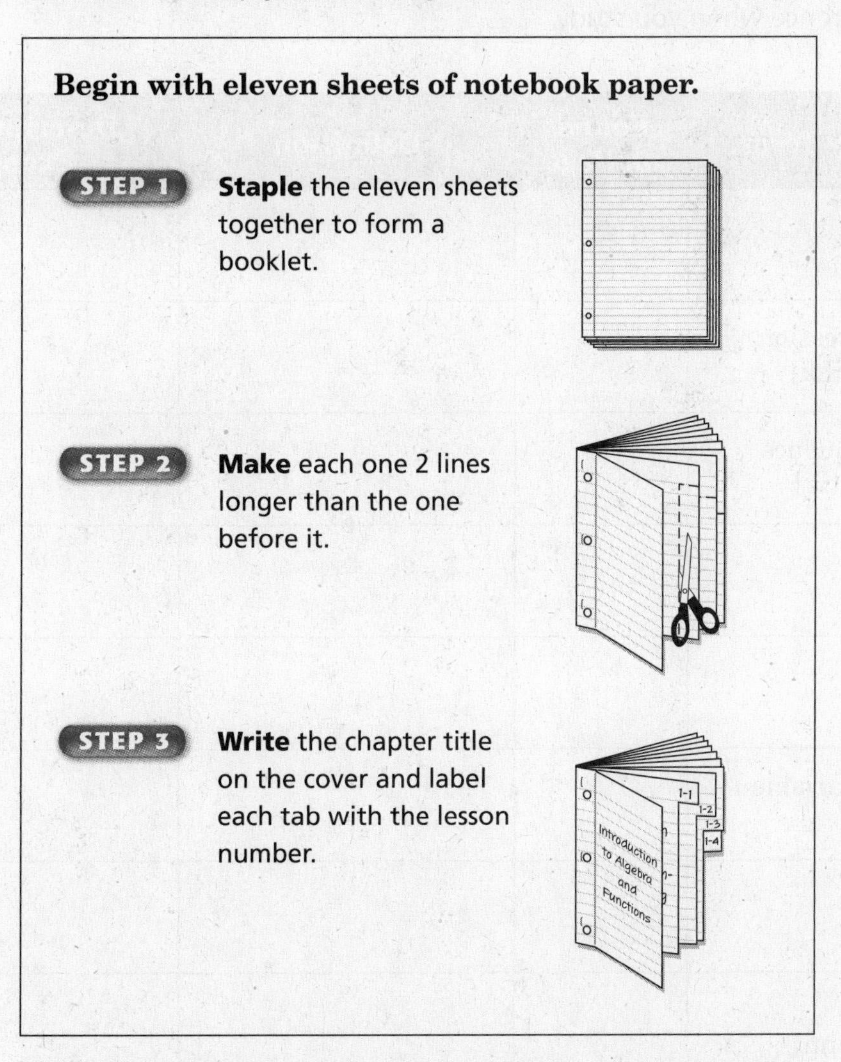

STEP 1 **Staple** the eleven sheets together to form a booklet.

STEP 2 **Make** each one 2 lines longer than the one before it.

STEP 3 **Write** the chapter title on the cover and label each tab with the lesson number.

NOTE-TAKING TIP: When taking notes, it is often a good idea to write a summary of the lesson in your own words. Be sure to paraphrase key points.

BUILD YOUR VOCABULARY

This is an alphabetical list of new vocabulary terms you will learn in Chapter 1. As you complete the study notes for the chapter, you will see Build Your Vocabulary reminders to complete each term's definition or description on these pages. Remember to add the textbook page number in the second column for reference when you study.

Vocabulary Term	Found on Page	Definition	Description or Example
algebra			
algebraic expression [al-juh-BRAY-ihk]			
arithmetic sequence [air-ith-MEH-tik]			
base			
coefficient			
defining the variable			
domain			
equation [ih-KWAY-zhuhn]			
equivalent expression			
evaluate			
exponent			

Vocabulary Term	Found on Page	Definition	Description or Example
factors			
function			
function rule			
numerical expression			
order of operations			
perfect square			
powers			
radical sign			
range			
sequence			
solution			
square			
square root			
term			
variable			

1–1 A Plan for Problem Solving

MAIN IDEA

- Solve problems using the four-step plan.

EXAMPLE Use the Four-Step Plan

1 SPENDING **A can of soda holds 12 fluid ounces. A 2-liter bottle holds about 67 fluid ounces. If a pack of six cans costs the same as a 2-liter bottle, which is the better buy?**

UNDERSTAND *What are you trying to find?* You know the number of fluid ounces of soda in one can of soda. You need to know the number of fluid ounces of soda in a pack of six cans.

PLAN You can find the number of fluid ounces of soda in a pack of six cans by ______ the number of fluid ounces in one can by ______.

SOLVE $12 \times$ ______ $=$ ______

There are ______ fluid ounces of soda in a pack of six cans. The number of fluid ounces of soda in a 2-liter bottle is about ______. Therefore, the ______ is the better buy because you get more soda for the same price.

CHECK The answer makes sense based on the facts given in the problem.

FOLDABLES

ORGANIZE IT

Summarize the four-step problem-solving plan on the Lesson 1-1 page of your Foldable.

Check Your Progress **FIELD TRIP** The sixth-grade class at Meadow Middle School is taking a field trip to the local zoo. There will be 142 students plus 12 adults going on the trip. If each school bus can hold 48 people, how many buses will be needed for the field trip?

EXAMPLE Use a Strategy in the Four-Step Plan

2 POPULATION For every 100,000 people in the United States, there are 5,750 radios. For every 100,000 people in Canada, there are 323 radios. Suppose Sheamus lives in Des Moines, Iowa, and Alex lives in Windsor, Ontario. Both cities have about 200,000 residents. About how many more radios are there in Sheamus's city than in Alex's city?

UNDERSTAND You know the approximate number of radios per 100,000 people in both Sheamus's city and Alex's city.

PLAN You can find the approximate number of radios in each city by ______ the estimate per 100,000 people by two to get an estimate per 200,000 people. Then, ______ to find how many more radios there are in Des Moines than in Windsor.

SOLVE Des Moines: 5,750 × 2 = ______

Windsor: 323 × 2 = ______

______ – ______ = ______

So, Des Moines has about ______ more radios than Windsor.

CHECK Based on the information given in the problem, the answer seems to be reasonable.

KEY CONCEPTS

Problem-Solving Strategies

- guess and check
- look for a pattern
- make an organized list
- draw a diagram
- act it out
- solve a simpler problem
- use a graph
- work backward
- eliminate possibilities
- estimate reasonable answers
- use logical reasoning
- make a model

Check Your Progress **READING** Ben borrows a 500-page book from the library. On the first day, he reads 24 pages. On the second day, he reads 39 pages and on the third day he reads 54 pages. If Ben follows the same pattern of number of pages read for seven days, will he have finished the book at the end of the week?

1–2 Powers and Exponents

BUILD YOUR VOCABULARY (pages 2–3)

Two or more numbers that are multiplied together to form a ______ are called **factors**.

The **exponent** tells how many times the base is used as a ______.

The **base** is the common ______.

Numbers expressed using ______ are called **powers**.

Five to the ______ power is five **squared**.

Four to the ______ power is four **cubed**.

MAIN IDEA

- Use powers and exponents.

FOLDABLES

ORGANIZE IT

On the Lesson 1-2 page of your Foldable, explain the difference between the terms power and exponent.

EXAMPLES Write Powers as Products

Write each power as a product of the same factor.

1 8^4

Eight is used as a factor ______ times. $8^4 =$ ______

2 4^6

______ is used as a factor six times. $4^6 =$ ______

Check Your Progress **Write each power as a product of the same factor.**

a. 3^6

b. 7^3

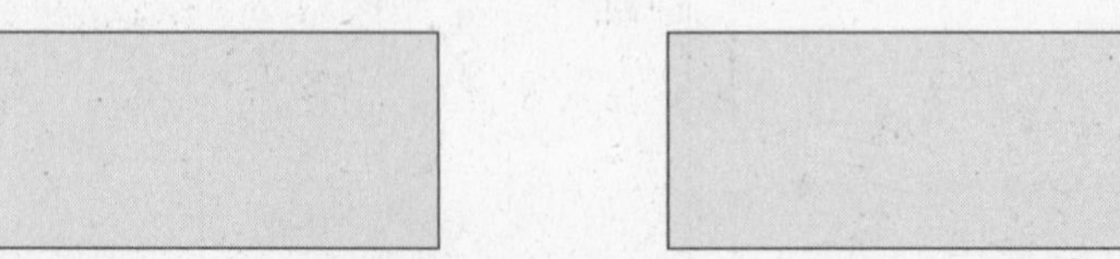

BUILD YOUR VOCABULARY (pages 2–3)

You can **evaluate**, or find the ______ of, ______ by multiplying the factors.

Numbers written ______ are in **standard form**.

Numbers written ______ are in **exponential form**.

WRITE IT

Explain how you would use a calculator to evaluate a power.

______
______
______
______
______

EXAMPLES Write Powers in Standard Form

Evaluate each expression.

3 8^3 = ______ = ______

4 6^4 = ______ = ______

Check Your Progress **Evaluate each expression.**

a. 4^4 **b.** 5^5

EXAMPLE Write Numbers in Exponential Form

5 **Write 9 • 9 • 9 • 9 • 9 • 9 in exponential form.**

9 is the ______. It is used as a factor ______ times.

So, the exponent is ______.

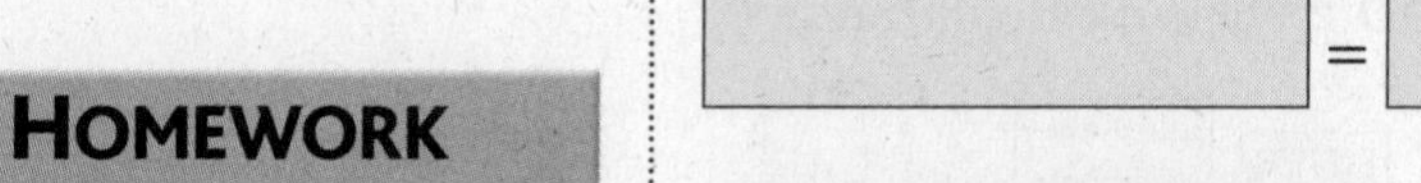

______ = ______

Check Your Progress Write 3 • 3 • 3 • 3 • 3 in exponential form.

HOMEWORK ASSIGNMENT

Page(s):

Exercises:

1–3 Squares and Square Roots

MAIN IDEA

- Find squares of numbers and square roots of perfect squares.

BUILD YOUR VOCABULARY (pages 2–3)

The ______ of a number and ______ is the **square** of the number.

Perfect squares like 9, 16, and 225 are squares of ______ numbers.

The ______ multiplied to form perfect squares are called **square roots**.

A **radical sign**, $\sqrt{}$, is the symbol used to indicate the positive ______ of a number.

FOLDABLES

ORGANIZE IT

On the Lesson 1-3 page of your Foldable, explain in words and symbols how you find squares of numbers and square roots of perfect squares.

EXAMPLES Find Squares of Numbers

1 **Find the square of 5.**

Multiply 5 by ______.

______ · ______ = 25

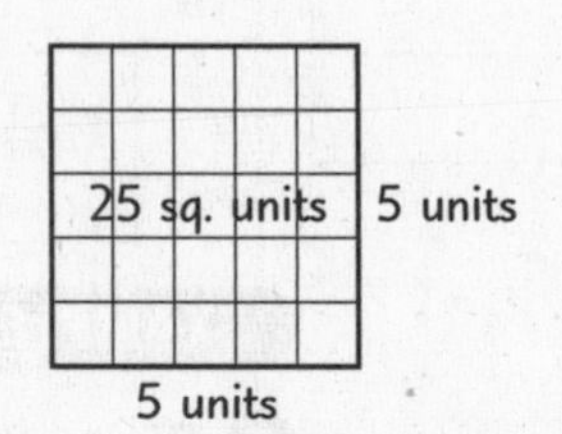

2 **Find the square of 19.**

METHOD 1 Use paper and pencil.

______ · ______ = ______

METHOD 2 Use a calculator.

Check Your Progress **Find the square of each number.**

a. 7

b. 21

KEY CONCEPT

Square Root A square root of a number is one of its two equal factors.

EXAMPLES Find Square Roots

3 Find $\sqrt{36}$.

What number times itself is 36?

$\square \cdot \square = 36$, so $\sqrt{36} = \square$.

4 Find $\sqrt{676}$.

Use a calculator.

2nd x^2 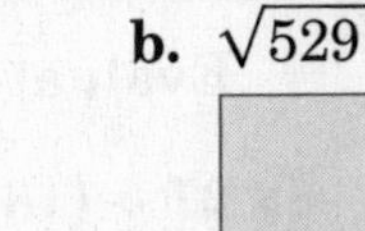 ENTER

So, $\sqrt{676} = \square$.

Check Your Progress **Find each square root.**

a. $\sqrt{64}$

b. $\sqrt{529}$

5 GAMES **A checkerboard is a square with an area of 1,225 square centimeters. What are the dimensions of the checkerboard?**

The checkerboard is a square. By finding the square root of the area, 1,225, you find the length of one side.

2nd x^2 ENTER

Use a calculator.

The dimensions of the checkerboard are $\square$ cm by $\square$ cm.

Check Your Progress **GARDENING** Kyle is planting a new garden that is a square with an area of 42.25 square feet. What are the dimensions of Kyle's garden?

2nd x^2 ENTER

HOMEWORK ASSIGNMENT

Page(s):

Exercises:

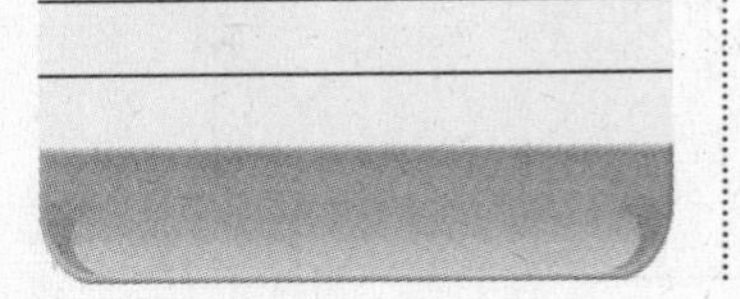

1–4 Order of Operations

MAIN IDEA

- Evaluate expressions using the order of operations.

BUILD YOUR VOCABULARY (pages 2–3)

The expressions $4 \cdot 6 - (5 + 7)$ and $8 \cdot (9 - 3) + 4$ are ______ expressions.

Order of operations are ______ that ensure that numerical expressions have only one value.

KEY CONCEPT

Order of Operations

1. Evaluate the expressions inside grouping symbols.
2. Evaluate all powers.
3. Multiply and divide in order from left to right.
4. Add and subtract in order from left to right.

FOLDABLES Be sure to include the order of operations on the Lesson 1-4 page of your Foldable.

EXAMPLES Evaluate Expressions

Evaluate each expression.

1 $27 - (18 + 2)$

$27 - (18 + 2) = 27 -$ ______ Add first since 18 + 2 is in parentheses.

$=$ ______ Subtract 20 from 27.

2 $15 + 5 \cdot 3 - 2$

$15 + 5 \cdot 3 - 2 = 15 +$ ______ $- 2$ Multiply 5 and 3.

$=$ ______ $- 2$ Add 15 and 15.

$=$ ______ Subtract 2 from 30.

Check Your Progress **Evaluate each expression.**

a. $45 - (26 + 3)$

b. $32 - 3 \cdot 7 + 4$

EXAMPLES Use Order of Operations

Evaluate each expression.

3 $12 \times 3 - 2^2$

$12 \times 3 - 2^2 = 12 \times 3 - \square$ Find the value of 2^2.

$= \square - 4$ Multiply 12 and 3.

$= \square$ Subtract 4 from 36.

4 $28 \div (3 - 1)^2$

$28 \div (3 - 1)^2 = 28 \div \square$ Subtract 1 from 3 inside the parentheses.

$= 28 \div \square$ Find the value of 2^2.

$= \square$ Divide.

REMEMBER IT

If an exponent lies outside of grouping symbols, complete the operations within the grouping symbols before applying the power.

Check Your Progress **Evaluate each expression.**

a. $9 \times 5 + 3^2$

b. $36 \div (14 - 11)^2$

EXAMPLE Evaluate an Expression

5 MONEY **Julian is buying one box of favors, one box of balloons, and three rolls of crepe paper. What is the total cost?**

Item	Quantity	Unit Cost
crepe paper	3 rolls	$2
favors	1 box	$7
balloons	1 box	$5

$1 \times 7 + 1 \times 5 + 3 \times 2 = 7 + \square + 6$ or 18

The total cost is $\square$.

Check Your Progress What is the total cost of two boxes of favors, two boxes of balloons, and six rolls of crepe paper?

HOMEWORK ASSIGNMENT

Page(s):

Exercises:

1–5 Problem-Solving Investigation: Guess and Check

EXAMPLE Use Guess and Check Strategy

MAIN IDEA

- Solve problems using the guess and check strategy.

CONCESSIONS **The concession stand at the school play sold lemonade for \$0.50 and cookies for \$0.25. They sold 7 more lemonades than cookies, and they made a total of \$39.50. How many lemonades and cookies were sold?**

UNDERSTAND You know the cost of each lemonade and cookie. You know the total amount made and that they sold ______ more lemonades than cookies. You need to know how many lemonades and cookies were sold.

PLAN Make a guess and check it. Adjust the guess until you get the correct answer.

SOLVE Make a guess.

14 cookies, 21 lemonades $0.25(14) + 0.50(21)$

This guess is too ______. $=$ ______

50 cookies, 57 lemonades $0.25(50) + 0.50(57)$

This guess is too ______. $=$ ______

48 cookies, 55 lemonades $0.25(48) + 0.50(55)$

$=$ ______

CHECK 48 cookies cost \$12 and 55 lemonades cost \$27.50. Since \$12 + \$27.50 = \$39.50 and 55 is 7 more than 48, the guess is correct.

Check Your Progress **ZOO** A total of 122 adults and children went to the zoo. Adult tickets cost \$6.50 and children's tickets cost \$3.75. If the total cost of the tickets was \$597.75, how many adults and children went to the zoo?

HOMEWORK ASSIGNMENT

Page(s):

Exercises:

1–6 Algebra: Variables and Expressions

BUILD YOUR VOCABULARY (pages 2–3)

MAIN IDEA

- Evaluate simple algebraic expressions.

You can use a ______, or **variable**, in an expression.

The expression $7 + n$ is called an ______ expression.

The branch of mathematics that involves expressions with ______ is called **algebra**.

The ______ factor of a term that contains a variable is called a **coefficient**.

EXAMPLES Evaluate Expressions

1 **Evaluate $t - 4$ if $t = 6$.**

$t - 4 = 6 - $ ______ Replace t with ______.

$=$ ______ Subtract.

2 **Evaluate $5x + 3y$ if $x = 7$ and $y = 9$.**

$5x + 3y = 5 \cdot$ ______ $+ 3 \cdot$ ______ Replace x with ______ and ______ with 9.

$=$ ______ $+$ ______ Do all multiplications first.

$=$ ______ Add ______ and 27.

3 **Evaluate $5 + a^2$ if $a = 5$.**

$5 + a^2 = 5 + 5^2$ Replace a with ______.

$= 5 +$ ______ Evaluate the ______.

$=$ ______ Add.

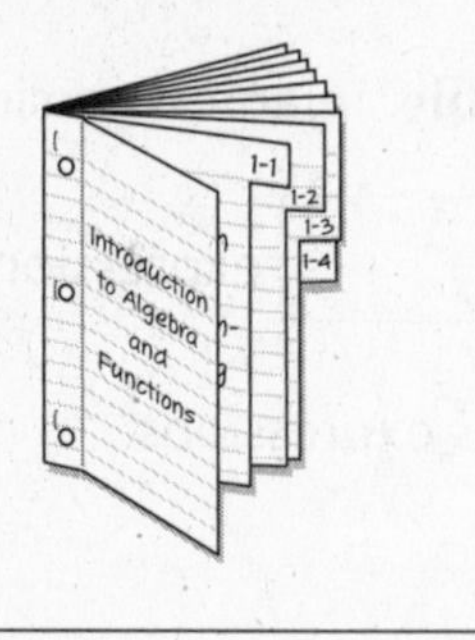

FOLDABLES

ORGANIZE IT

Record and evaluate an example of a simple algebraic expression on the Lesson 1-6 page of your Foldable.

Check Your Progress **Evaluate each expression.**

a. $7 + m$ if $m = 4$.

b. $4a - 2b$ if $a = 9$ and $b = 6$.

c. $24 - s^2$ if $s = 3$.

EXAMPLE Evaluate an Expression

4 TEMPERATURE The formula for rewriting a Fahrenheit temperature as a Celsius temperature is $\frac{5(F - 32)}{9}$, where F equals the temperature in degrees Fahrenheit. Find the Celsius equivalent of 99°F.

$\frac{5(F - 32)}{9} = \frac{5(99 - 32)}{9}$ Replace F with 99.

$= \frac{5(67)}{9} = \frac{335}{9}$ Subtract ______ from 99 and multiply.

$\approx$ ______ Divide 335 by 9.

The Celsius equivalent of 99°F is about 37.2°C.

Check Your Progress **BOWLING** David's cost for bowling can be described by the formula $1.75 + 2.5g$, where g is the number of games David bowls. Find the total cost of bowling if David bowls 3 games.

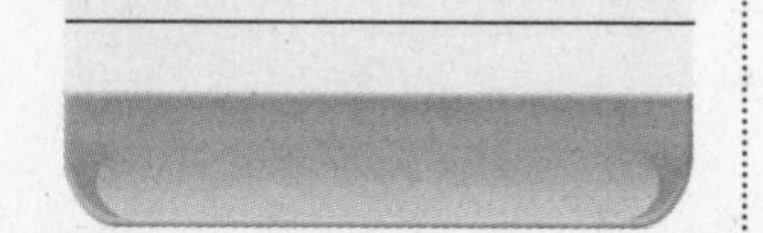

HOMEWORK ASSIGNMENT

Page(s):

Exercises:

1–7 Algebra: Equations

MAIN IDEA

- Write and solve equations using mental math.

BUILD YOUR VOCABULARY (pages 2–3)

An **equation** is a ______ in mathematics that contains an equals sign.

The **solution** of an equation is a number that makes the sentence ______.

The process of finding a ______ is called **solving an equation.**

When you choose a ______ to represent one of the unknowns in an equation, you are **defining the variable.**

FOLDABLES

ORGANIZE IT

On the Lesson 1-7 page of your Foldable, record and solve an example of an algebraic equations.

EXAMPLE Solve an Equation Mentally

1 Solve $p - 14 = 5$ mentally.

$p - 14 = 5$ Write the equation.

______ $- 14 = 5$ You know that 19 – 14 is ______.

______ $= 5$ Simplify.

The solution is ______.

Check Your Progress Solve $p - 6 = 11$ mentally.

EXAMPLE

2 TEST EXAMPLE A store sells pumpkins for $2 per pound. Paul has $18. Use the equation $2x = 18$ to find how large a pumpkin Paul can buy with $18.

A 6 lb **B** 7 lb **C** 8 lb **D** 9 lb

Read the Item

Solve  to find how many pounds the pumpkin can weigh.

Solve the Item

Write the equation.

$2 \cdot$ ___ $= 18$ You know that $2 \cdot 9$ is 18.

Paul can buy a pumpkin as large as ___ pounds.

The answer is ___.

Check Your Progress A store sells notebooks for $3 each. Stephanie has $15. Use the equation $3x = 15$ to find how many notebooks she can buy with $15.

F 4 **G** 5 **H** 6 **J** 7

EXAMPLE Write an Equation to Solve a Problem

REVIEW IT

Explain how to add a decimal and a whole number. *(Prerequisite Skill)*

3 ENTERTAINMENT **An adult paid $18.50 for herself and two students to see a movie. If the two student tickets cost $11 together, what is the cost of the adult ticket?**

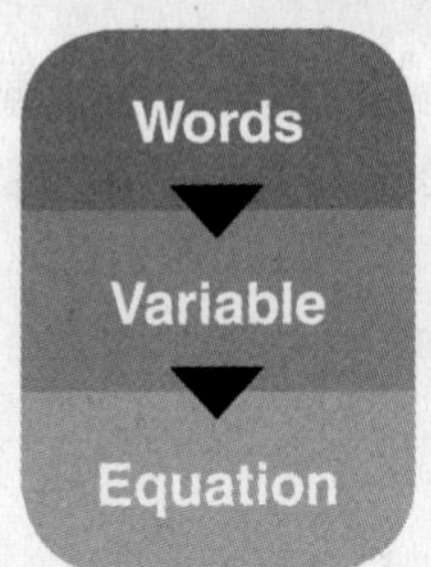

The cost of one adult ticket and two student tickets is $18.50.

Let a represent the cost of an adult movie ticket.

$a + 11 = 18.50$

$a + 11 = 18.50$	Write the equation.
______ $+ 11 = 18.50$	Replace a with ______ to make the equation true.
______ $= 18.50$	Simplify.

The number ______ is the solution of the equation. So, the cost of an adult movie ticket is ______.

Check Your Progress **ICE CREAM** Julie spends $9.50 at the ice cream parlor. She buys a hot fudge sundae for herself and ice cream cones for each of the three friends who are with her. Find the cost of Julie's sundae if the three ice cream cones together cost $6.30.

HOMEWORK ASSIGNMENT

Page(s):

Exercises:

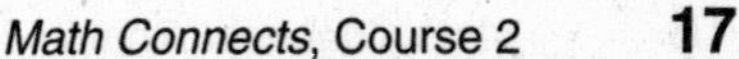

1–8 Algebra: Properties

BUILD YOUR VOCABULARY (pages 2–3)

The expressions 5($9 + $2) and 5($9) + 5($2) are **equivalent expressions** because they have the ______ value.

MAIN IDEA

- Use Commutative, Associative, Identity, and Distributive properties to solve problems.

EXAMPLES Use the Distributive Property

Use the Distributive Property to rewrite each expression. Then evaluate it.

1 $8(5 + 7)$

$8(5 + 7) = 8 \cdot \_\_\_ + 8 \cdot \_\_\_$

$= \_\_\_ + \_\_\_$ Multiply.

$= \_\_\_$ Add.

KEY CONCEPT

Distributive Property To multiply a sum by a number, multiply each addend of the sum by the number outside the parentheses.

2 $6(9) + 6(2)$

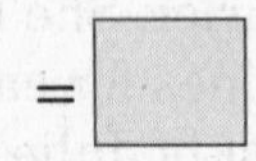

$6(9) + 6(2) = \_\_\_ + \_\_\_$ Multiply.

$= \_\_\_$ Add.

FOLDABLES

ORGANIZE IT

On the Lesson 1-8 page your Foldable, be sure to include examples showing the addition and multiplication properties.

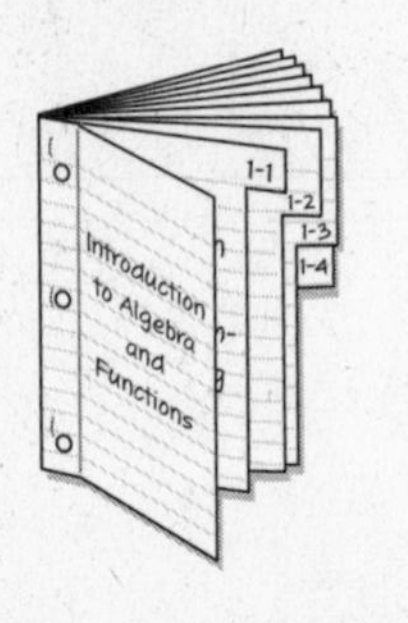

Check Your Progress **Use the Distributive Property to evaluate each expression.**

a. $4(6 + 3)$

b. $(5 + 3)7$

KEY CONCEPTS

Commutative Property The order in which two numbers are added or multiplied does not change their sum or product.

Associative Property The way in which three numbers are grouped when they are added or multiplied does not change their sum or product.

Identity Property The sum of an addend and zero is the addend. The product of a factor and one is the factor.

EXAMPLE

3 VACATIONS **Mr. Harmon has budgeted $150 per day for his hotel and meals during his vacation. If he plans to spend six days on vacation, how much will he spend?**

$6(150) = 6\left(100 + \_\_\_\_\right)$	150 = 100 + 50.
$= \_\_\_\_(100) + \_\_\_\_(50)$	Distributive Property
$= 600 + \_\_\_\_$ or 900	Multiply, then add.

Mr. Harmon will spend about ______ on a six-day vacation.

Check Your Progress **COOKIES** Heidi sold cookies for \$2.50 per box for a fundraiser. If she sold 60 boxes of cookies, how much money did she raise?

______

BUILD YOUR VOCABULARY (pages 2–3)

Properties are statements that are ______ for all numbers.

EXAMPLE Identify Properties

4 **Find 5 • 13 • 20 mentally. Justify each step.**

$5 \cdot 13 \cdot 20 = 5 \cdot \_\_\_\_ \cdot \_\_\_\_$	Communtative Property of Multiplication
$= \left(\_\_\_\_ \cdot 20\right) \cdot 13$	Associative Property of Multiplication
$= \_\_\_\_ \cdot 13$ or ______	Multiply 100 and 13 Mentally.

Check Your Progress Name the property shown by the statement $4 + (6 + 2) = (4 + 6) + 2$.

______

HOMEWORK ASSIGNMENT

Page(s):

Exercises:

1–9 Algebra: Arithmetic Sequences

MAIN IDEA

- Describe the relationships and extend terms in arithmetic sequences.

BUILD YOUR VOCABULARY (pages 2–3)

A **sequence** is an ______ list of ______.

Each number in a ______ is called a **term**.

In an **arithmetic sequence**, each term is found by ______ the same number to the ______ term.

FOLDABLES

ORGANIZE IT

Write an example of an arithmetic and a geometric sequence on the Lesson 1-9 page of your Foldable.

EXAMPLES Describe Patterns in Sequences

Describe the relationship between the terms in each arithmetic sequence. Then write the next three terms in the sequence.

1 **7, 11, 15, 19, ...**

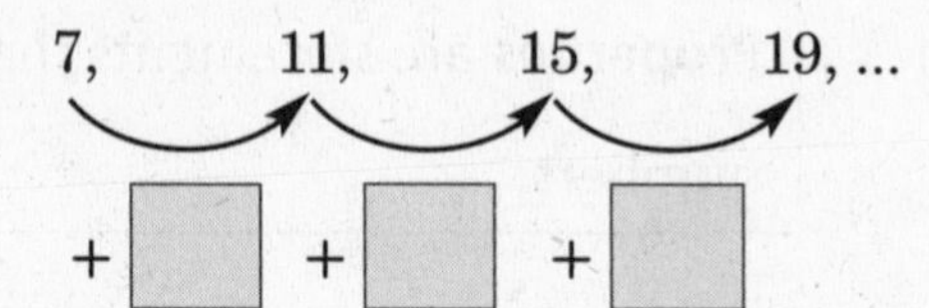

Each term is found by ______ 4 to the previous term.

Continue the pattern to find the next three terms.

$19 + 4 =$ ______ $\quad 23 + 4 =$ ______ $\quad 27 + 4 =$ ______

The next three terms are 23, 27, and 31.

2 **0.1, 0.5, 0.9, 1.3, ...**

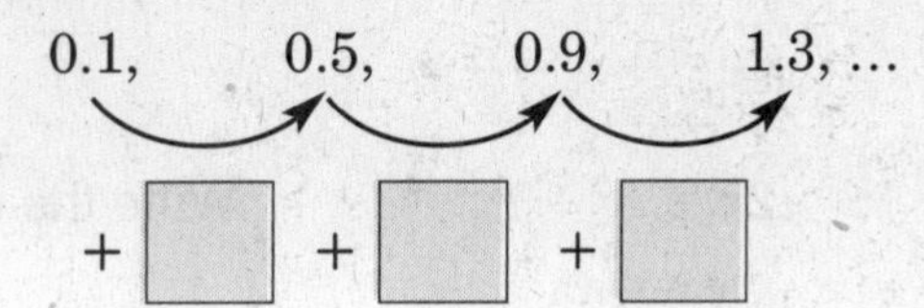

Each term is found by adding ☐ to the previous term.

Continue the pattern to find the next three terms.

1.3 + ☐ = 1.7   1.7 + ☐ = ☐   2.1 + 0.4 = ☐

The next three terms are 1.7, 2.1, and 2.5.

Check Your Progress **Describe the relationship between the terms in each arithmetic sequence. Then write the next three terms in the sequence.**

a. 13, 24, 35, 46, ...

b. 0.6, 1.5, 2.4, 3.3, ...

WRITE IT

In your own words, explain how to determine the pattern in a sequence.

EXAMPLE Use a Table

3 **EXERCISE Mehmet started a new exercise routine. The first day, he did 2 sit-ups. Each day after that, he did 2 more sit-ups than the previous day. If he continues this pattern, how many sit-ups will he do on the tenth day?**

Position	Operation	Value of Term
1		2
2	$2 \cdot 2$	
	$3 \cdot 2$	6
d	$d \cdot 2$	$2d$

(continued on the next page)

Each term is 2 times its position number. So, the expression is ______.

$2n$ Write the expression.

$2(\_\_\_) = 20$ Replace n with 10.

So, on the tenth day, Mehmet will do ______ sit-ups.

Check Your Progress **CONCERTS** The first row of a theater has 8 seats. Each additional row has eight more seats than the previous row. If this pattern continues, what algebraic expression can be used to find the number of seats in the 15th row? How many seats will be in the 15th row?

HOMEWORK ASSIGNMENT

Page(s):

Exercises:

1–10 Algebra: Equations and Functions

MAIN IDEA

- Make function tables and write equations.

BUILD YOUR VOCABULARY (pages 2–3)

A relationship where one thing depends on another is called a **function**.

The ______ performed on the input is given by the **function rule**.

You can organize the ______ numbers, ______ numbers, and the function rule in a **function table**.

The set of ______ values is called the **domain**.

The set of ______ values is called the **range**.

REMEMBER IT

When x and y are used in an equation, x usually represents the input and y usually represents the output.

EXAMPLE Make a Function Table

1 **Asha earns $6.00 an hour working at a grocery store. Make a function table that shows Asha's total earnings for working 1, 2, 3, and 4 hours.**

Input	Function	Output
Number of Hours	Multiply by 6	Total Earnings ($)
1		6
2	6 × 2	
	6 × 3	18
4		

Check Your Progress **MOVIE RENTAL** Dave goes to the video store to rent a movie. The cost per movie is $3.50. Make a function table that shows the amount Dave would pay for renting 1, 2, 3, and 4 movies.

EXAMPLES

2 READING Melanie read 14 pages of a detective novel each hour. Write an equation using two variables to show how many pages p she read in h hours.

Input	Function	Output
Number of Hours (h)	Multiply by 14	Number of Pages Read (p)
1	1×14	
2		28
	3×14	42
h		$14h$

Words: **number of pages read** equals ____ **pages** times **number of hours**

Variable: Let p represent the number of pages read.
Let ____ represent the number of hours.

Equation: $p =$ ____

3 READING **Use your equation from Example 2 to find how many pages Melanie read in 7 hours.**

[ ] Write the equation.

$p = 14$ ([ ]) Replace h with 7.

$p =$ [ ] Multiply.

Melanie read 98 pages in 7 hours.

Check Your Progress

a. **TRAVEL** Derrick drove 55 miles per hour to visit his grandmother. Write an equation using two variables to show how many miles m he drove in h hours.

b. **TRAVEL** Use your equation from above to find how many miles Derrick drove in 6 hours.

HOMEWORK ASSIGNMENT

Page(s):

Exercises:

CHAPTER 1

BRINGING IT ALL TOGETHER

STUDY GUIDE

FOLDABLES	VOCABULARY PUZZLEMAKER	BUILD YOUR VOCABULARY
Use your **Chapter 1 Foldable** to help you study for your chapter test.	To make a crossword puzzle, word search, or jumble puzzle of the vocabulary words in Chapter 1, go to: glencoe.com	You can use your completed **Vocabulary Builder** (*pages 2–3*) to help you solve the puzzle.

1-1 A Plan for Problem Solving

Underline the correct term to complete each sentence.

1. The (*Plan*, *Solve*) step is the step of the four-step plan in which you decide which strategy you will use to solve the problem.

2. According to the four-step plan, if your answer is not correct, you should (*estimate the answer*, *make a new plan and start again*).

3. Once you solve a problem, make sure your solution contains any appropriate (*strategies*, *units or labels*).

1-2 Powers and Exponents

Identify the exponent in each expression.

4. 5^8 ______ 5. 8^3 ______

Evaluate each expression.

6. 4^3 ______ 7. 8^5 ______

Complete the sentence.

8. Numbers written with exponents are in ______ form, whereas numbers written without exponents are in ______ form.

1-3 Squares and Square Roots

Complete each sentence.

9. The square of 3 means ____ × ____.

10. Nine units squared means 9 ____ with ____ of ____ unit each.

Find the square of each number.

11. 16 ____

12. 28 ____

Find the square root of each number.

13. $\sqrt{121}$ ____

14. $\sqrt{484}$ ____

1-4 Order of Operations

Evaluate each expression.

15. $9 + 18 \div 6$

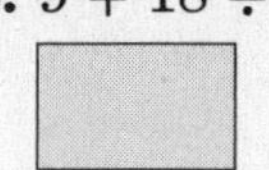

16. $(7 - 4)^2 \div 3$

17. $2 \times 4^2 \div 4 - 1$

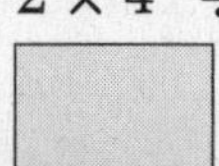

18. $8 + 2(9 - 5) - (2 \cdot 3)$

1-5 Problem-Solving Investigation: Guess and Check

Solve using the *guess and check* strategy.

19. MONEY Gary deposited $38 into his savings account every week for eight weeks. At the end of this time, the total amount in his account was $729. How much money did Gary have in his account before the deposits?

1-6

Algebra: Variables and Expressions

Evaluate each expression if $a = 5$ and $b = 6$.

20. $2a + 3b$

21. $\frac{ab}{5}$

22. $a^2 - 3b$

1-7

Algebra: Equations

Solve each equation mentally.

23. $5 + b = 12$

24. $h - 6 = 3$

25. $12 \cdot 4 = n$

26. $2 = \frac{x}{4}$

27. $9t = 54$

28. $35 \div c = 7$

1-8

Algebra: Properties

Match the statement with the property it shows.

29. $5 + (3 + 6) = (5 + 3) + 6$

30. $8 + 0 = 8$

31. $4(7 - 2) = 4(7) - 4(2)$

32. $10 + 9 = 9 + 10$

a. Distributive Property

b. Commutative Property of Addition

c. Associative Property of Addition

d. Identity Property of Addition

1-9

Algebra: Arithmetic Sequences

Complete the sentence.

33. In an arithmetic sequence, each term is found by ____ the same number to the previous term.

34. In a geometric sequence, each term is found by ____ the same number by the previous term.

What is the next term in each of the following sequences?

35. 1, 5, 25, … ____

36. 7, 10, 13, … ____

1-10

Algebra: Equations and Functions

37. Complete the function table. Identify the domain and range. Then graph the function.

x	$2x - 1$	y
-1		
0		
1		

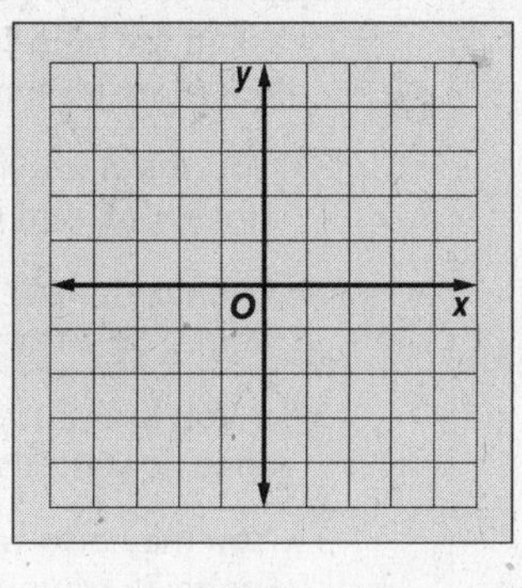

Domain = ____

Range = ____

ARE YOU READY FOR THE CHAPTER TEST?

Math Online

Visit **glencoe.com** to access your textbook, more examples, self-check quizzes, and practice tests to help you study the concepts in Chapter 1.

Check the one that applies. Suggestions to help you study are given with each item.

☐ **I completed the review of all or most lessons without using my notes or asking for help.**

- You are probably ready for the Chapter Test.
- You may want to take the Chapter 1 Practice Test on page 75 of your textbook as a final check.

☐ **I used my Foldables or Study Notebook to complete the review of all or most lessons.**

- You should complete the Chapter 1 Study Guide and Review on pages 70–74 of your textbook.
- If you are unsure of any concepts or skills, refer back to the specific lesson(s).
- You may want to take the Chapter 1 Practice Test on page 75 of your textbook.

☐ **I asked for help from someone else to complete the review of all or most lessons.**

- You should review the examples and concepts in your Study Notebook and Chapter 1 Foldables.
- Then complete the Chapter 1 Study Guide and Review on pages 70–74 of your textbook.
- If you are unsure of any concepts or skills, refer back to the specific lesson(s).
- You may also want to take the Chapter 1 Practice Test on page 75 of your textbook.

Student Signature

Parent/Guardian Signature

Teacher Signature

Integers

Use the instructions below to make a Foldable to help you organize your notes as you study the chapter. You will see Foldable reminders in the margin of this Interactive Study Notebook to help you in taking notes.

Begin with two sheets of $8\frac{1}{2}" \times 11"$ paper.

STEP 1 **Fold** one sheet in half from top to bottom. Cut along fold from edges to margin.

STEP 2 **Fold** the other sheet in half from top to bottom. Cut along fold between margins.

STEP 3 **Insert** first sheet through second sheet and align folds.

STEP 4 **Label** each page with a lesson number and title.

2-1
Integers and
Absolute Value

NOTE-TAKING TIPS: When you take notes, it is helpful to list ways in which the subject matter relates to daily life.

BUILD YOUR VOCABULARY

This is an alphabetical list of new vocabulary terms you will learn in Chapter 2. As you complete the study notes for the chapter, you will see Build Your Vocabulary reminders to complete each term's definition or description on these pages. Remember to add the textbook page number in the second column for reference when you study.

Vocabulary Term	Found on Page	Definition	Description or Example
absolute value			
additive inverse			
coordinate plane			
graph			
integer [IHN-tih-juhr]			
negative integer			
opposites			

Vocabulary Term	Found on Page	Definition	Description or Example
ordered pair			
origin			
positive integer			
quadrant			
x-axis			
x-coordinate			
y-axis			
y-coordinate			

2-1 Integers and Absolute Value

BUILD YOUR VOCABULARY (pages 32–33)

An **integer** is any ______ from the set {..., −4, ______, −2, −1, 0, 1, ______, 3, 4, ...}.

To **graph** a ______ on the number line, draw a point on the line at its ______.

Negative integers are integers ______ than zero.

Positive integers are integers ______ than zero.

MAIN IDEA

- Read and write integers, and find the absolute value of a number.

FOLDABLES

ORGANIZE IT

Under Lesson 2-1 in your notes, draw a number line and graph a few positive and negative integers. Then write a few real world situations that can be described by negative numbers.

2-1
Integers and Absolute Value

EXAMPLES Write Integers for Real-Life Situations

Write an integer for each situation.

1 a total rainfall of 2 inches below normal

Because it represents below normal, the integer is ______.

2 a seasonal snowfall of 3 inches above normal

Because it represents ______ normal, the integer is ______.

Check Your Progress **Write an integer for each situation.**

a. a total snowfall of 5 inches above normal

b. an average monthly temperature of 4 degrees below normal

BUILD YOUR VOCABULARY (pages 32–33)

The numbers ______ and 5 are the same ______ from 0, so −5 and 5 have the same **absolute value**.

EXAMPLE Graph Integers

3 Graph the set of integers {−1, 3, −2} on a number line.

Draw a number line. Then draw a ______ at the location of each integer.

Check Your Progress Graph the set of integers {−2, 1, −4} on a number line.

EXAMPLES Evaluate Expressions

KEY CONCEPT

Absolute Value The absolute value of an integer is the distance the number is from zero on a number line.

4 Evaluate the expression $|-5|$.

On the number line, the graph of −5 is 5 units from 0.

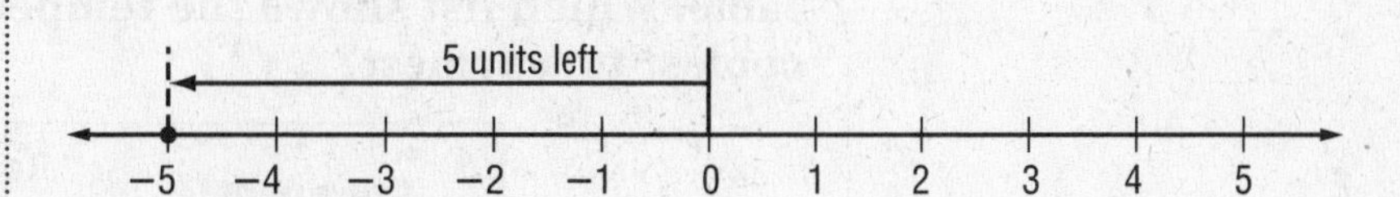

So, $|-5| =$ ______.

5 Evaluate the expression $|-4| - |-3|$.

$|-4| - |-3| =$ ______ − ______ $|-4| =$ ______, $|-3| =$ ______

$=$ ______   Subtract.

HOMEWORK ASSIGNMENT

Page(s):

Exercises:

Check Your Progress **Evaluate each expression.**

a. $|-9|$ ______   **b.** $|8| - |-5|$ ______

2-2 Comparing and Ordering Integers

MAIN IDEA

- Compare and order integers

EXAMPLE Compare Integers

1 **Replace the ● with < or > to make −9 ● −5 a true sentence.**

Graph each integer on a number line.

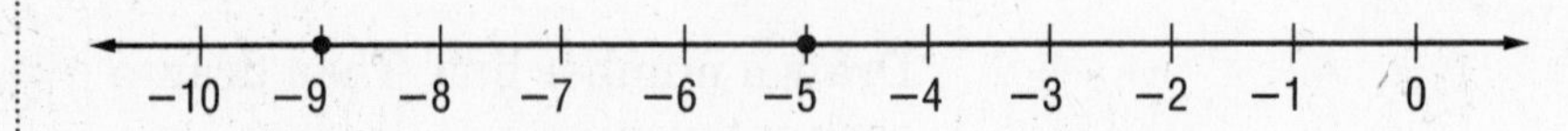

Since ______ is to the ______ of −5, −9 ______ −5.

Check Your Progress Replace the ● with < or > to make −3 ● −6 a true sentence.

EXAMPLE Order Integers

2 TEST EXAMPLE **The lowest temperatures in Europe, Greenland, Oceania, and Antarctica are listed in the table. Which list shows the temperatures in order from coolest to warmest?**

Continent	Record Low Temperature (°F)
Europe	−67
Greenland	−87
Oceania	14
Antarctica	−129

Source: *The World Almanac*

A −67, −87, 14, −129

B 14, −67, −87, −129

C −129, −87, −67, 14

D −67, −87, −129, 14

(continued on the next page)

ORGANIZE IT

Under Lesson 2-2 in your Foldable, explain how to compare integers. Be sure to include examples.

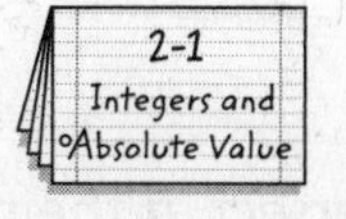

Read the Item

To order the integers, graph them on a number line.

Solve the Item

A G E O

−130 −120 −110 −100 −90 −80 −70 −60 −50 −40 −30 −20 −10 0 10 20

Order the integers from least to greatest by reading from left to right. The order from least to greatest is ______, ______, ______, ______. The answer is ______.

Check Your Progress **MULTIPLE CHOICE** The lowest temperatures on a given day in four cities in the United States are listed in the table. Which of the following lists the temperatures in order from coolest to warmest?

City	Low Temperature
Portland, OR	−12
New York City, NY	−6
Denver, CO	7
Newport, RI	−3

F −3, −6, 7, 12

G −12, −6, −3, 7

H −12, 7, −6, −3

J −3, −6, 7, −12

Page(s):

Exercises:

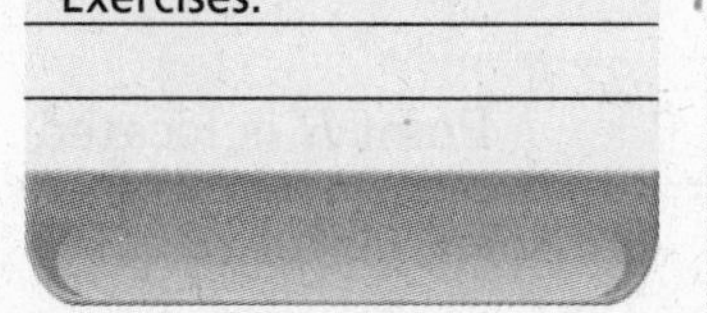

2–3 The Coordinate Plane

MAIN IDEA

- Graph points on a coordinate plane.

BUILD YOUR VOCABULARY (pages 32–33)

A **coordinate plane** is a plane in which a ______ number line and a vertical number line intersect at their zero points.

The ______ number line of a coordinate plane is called the ***x*-axis**.

The ______ number line of a coordinate plane is called the ***y*-axis**.

The **origin** is the point at which the number lines intersect in a coordinate grid.

An **ordered pair** is a pair of numbers such as (5, –2) used to locate a point in the coordinate plane. The ***x*-coordinate** is the ______ number. The ***y*-coordinate** is the ______ number.

FOLDABLES

ORGANIZE IT

Under Lesson 2-3 in your Foldable, record and define key terms about the coordinate system and give examples of each.

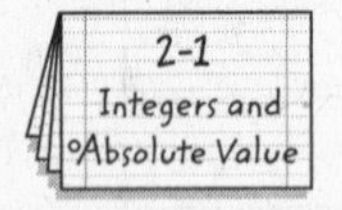

EXAMPLE Naming Points Using Ordered Pairs

1 Write the ordered pair that corresponds to point *R*. Then state the quadrant in which the point is located.

- Start at the origin.
- Move ______ to find the x-coordinate of point R, which is ______.
- Move up to find the ______, which is ______.

So, the ordered pair for point R is ______. Point R is located in Quadrant ______.

WRITE IT

When no numbers are shown on the *x*- or *y*-axis, how long is each interval?

____

____

____

____

Check Your Progress Write the ordered pair that names point M. Then name the quadrant in which the point is located.

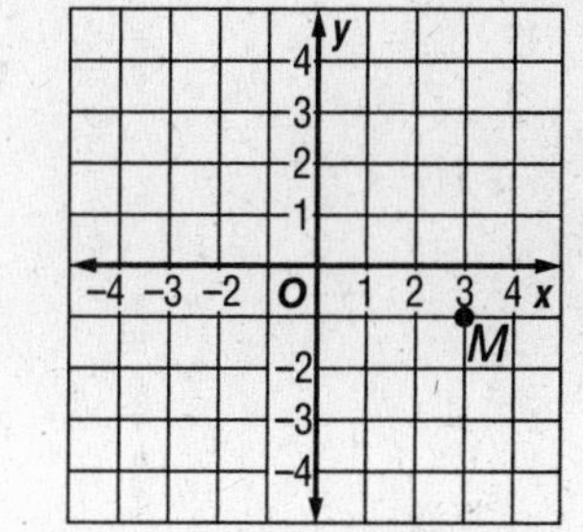

EXAMPLES Graph an Ordered Pair

2 Graph and label the point $M(3, 5)$.

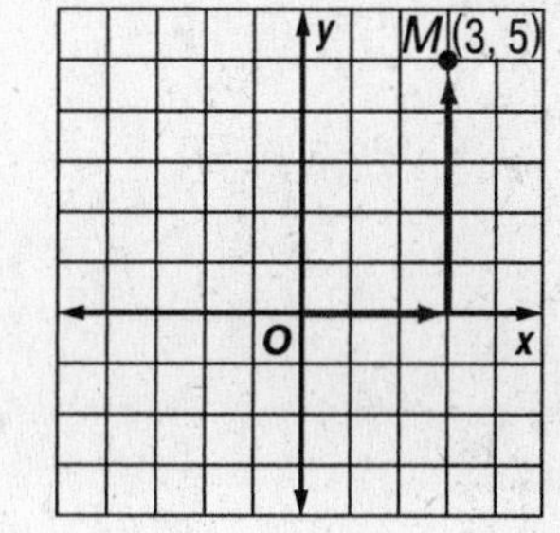

- Draw a coordinate plane.

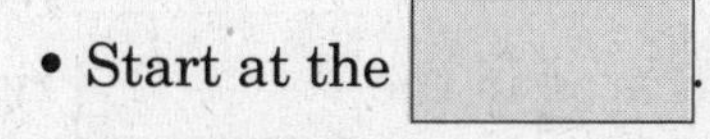

- Start at the ______.

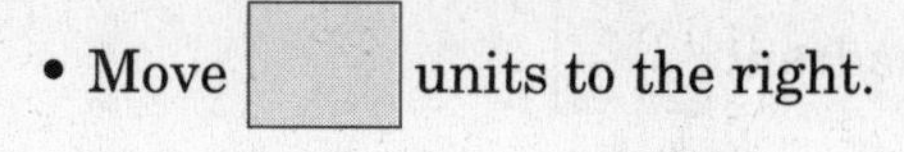

- Move ______ units to the right.

  Then move 5 units ______.

- Draw a dot and label it M ______.

Check Your Progress Graph and label the point $G(-2, -4)$.

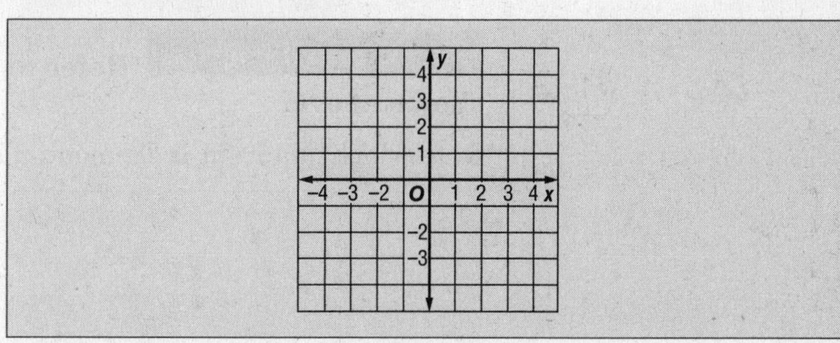

BUILD YOUR VOCABULARY (pages 32–33)

The coordinate plane is separated into ______ sections called **quadrants.**

Quadrant II (−, +)	Quadrant I (+, +)
Quadrant III (−, −)	Quadrant IV (+, −)

EXAMPLES Identify Quadrants

3 GEOGRAPHY **Use the map of Utah shown below.**

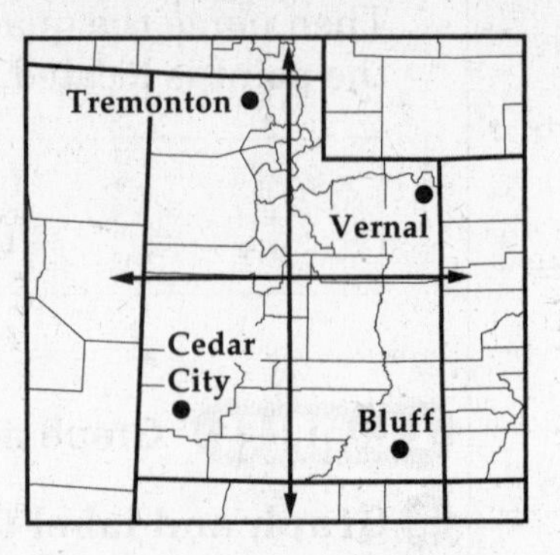

In which quadrant is Vernal located?

Vernal is located in the ______ right quadrant.

Quadrant ______.

4 **Which of the cities labeled on the map of Utah is located in quadrant IV?**

Quadrant ______ is the bottom right quadrant. So, ______ is in Quadrant IV.

Check Your Progress **Refer to the map of Utah shown above.**

a. In which quadrant is Tremonton located?

b. Which of the cities labeled on the map of Utah shown above is located in Quadrant III?

HOMEWORK ASSIGNMENT

Page(s):

Exercises:

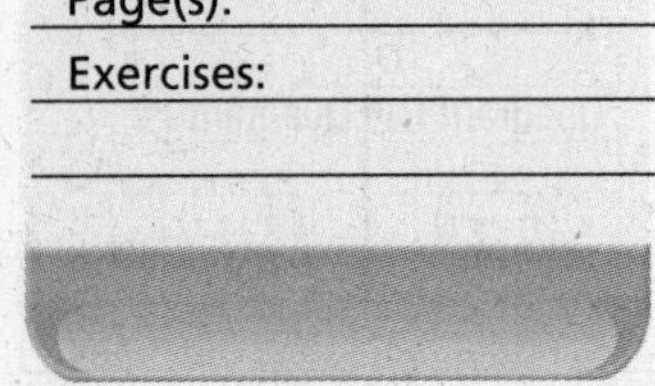

2-4 Adding Integers

EXAMPLES Add Integers with the Same Sign

MAIN IDEA

- Add integers.

1 Find −6 + (−3).

Use a number line.

- Start at ____.
- Move 6 units ____ to show −6.
- From there, move ____ units left to show ____.

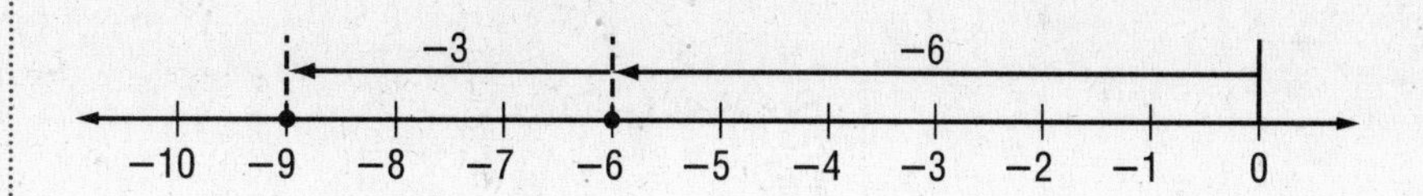

So, −6 + (−3) = ____.

KEY CONCEPTS

Adding Integers with the Same Sign The sum of two positive integers is positive. The sum of two negative integers is negative.

Additive Inverse Property The sum of any number and its additive inverse is 0.

2 Find −34 + (−21).

−34 + (−21) = ____ Both integers are negative, so the sum is ____.

Check Your Progress **Find each sum.**

a. −5 + (−2)

b. −27 + (−19)

BUILD YOUR VOCABULARY (pages 32–33)

The integers 5 and −5 are called **opposites** of each other because they are the same distance from 0, but on

 sides of 0.

Two ____ that are ____ are also called **additive inverses**.

KEY CONCEPT

Adding Integers with Different Signs To add integers with different signs, subtract their absolute values. The sum is:

- positive if the positive integer has the greater absolute value.
- negative if the negative integer has the greater absolute value.

EXAMPLES Add Integers with Different Signs

3 Find 8 + (−7).

Use a number line.

Start at ______.

Move ______ units right.

Then move ______ units left.

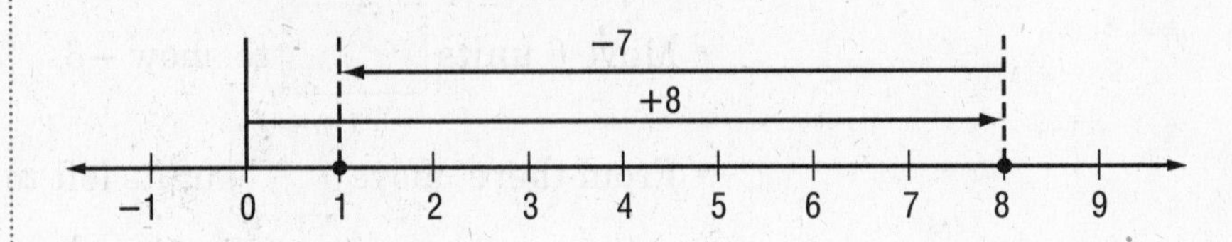

So, 8 + (−7) = ______.

4 Find −5 + 4.

Use a number line.

Start at ______.

Move ______ units left.

Then move 4 units ______.

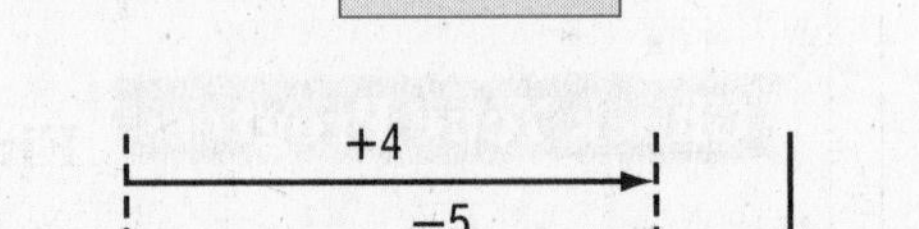

So, −5 + ______ = −1

FOLDABLES

ORGANIZE IT

Summarize the steps for adding integers. Be sure to include examples.

Check Your Progress **Add.**

a. 6 + (−2)

b. −3 + 5

EXAMPLES Add Integers with Different Signs

5 Find $2 + (-7)$.

$2 + (-7) =$ ______. Subtract absolute values; $7 - 2 = 5$. Since ______ has the greater absolute value, the sum is ______.

6 Find $-9 + 6$.

$-9 + 6 =$ ______ ______ the absolute values; $9 - 6 = 3$. Since -9 has the ______ absolute value, the sum is negative.

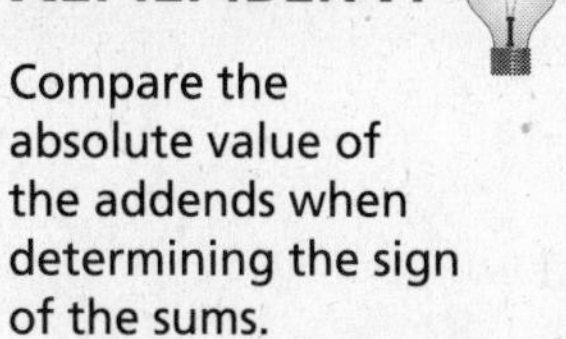

REMEMBER IT Compare the absolute value of the addends when determining the sign of the sums.

Check Your Progress **Add.**

a. $5 + (-9)$ **b.** $7 + (-3)$

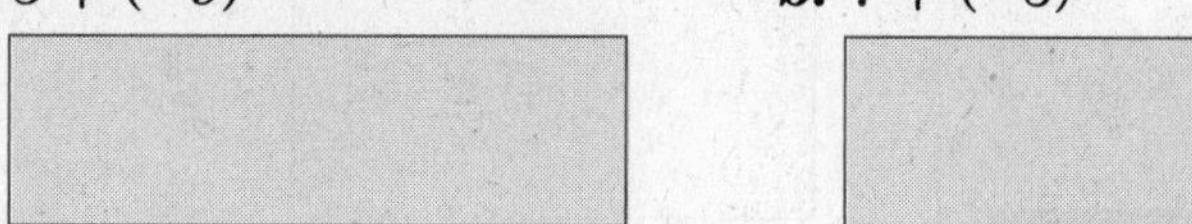

EXAMPLE Use the Additive Inverse Property

7 Find $11 + (-4) + (-11)$.

$11 + (-4) + (-11) = 11 + (-11) + (-4)$	Commutative Property (+)
$=$ ______ $+ (-4)$	Additive Inverse Property
$= -4$	Identity Property (+)

Check Your Progress Find $5 + (-11) + (-5)$.

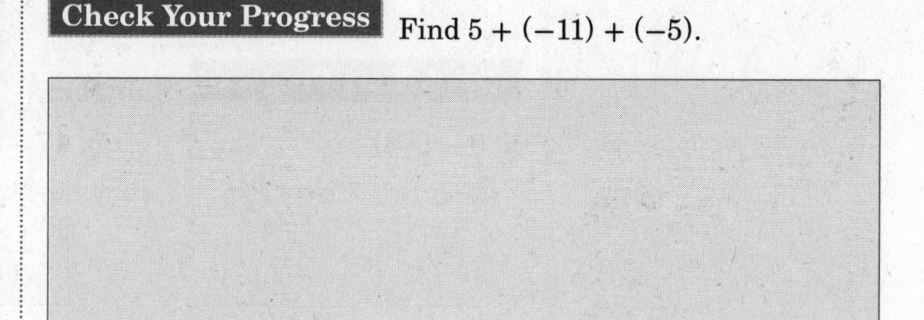

HOMEWORK ASSIGNMENT

Page(s):

Exercises:

2–5 Subtracting Integers

MAIN IDEA

- Subtract integers.

EXAMPLES Subtract Positive Integers

1 Find 2 – 15.

$2 - 15 = 2 + (-15)$ To subtract 15, add ______.

$= -13$ Simplify.

2 Find –13 – 8.

$-13 - 8 = -13 +$ ______ To subtract 8, add ______.

$= -21$ Simplify.

Check Your Progress **Subtract.**

a. $13 - 21$ **b.** $-9 - 11$

KEY CONCEPT

Subtracting Integers To subtract an integer, add its opposite.

FOLDABLES Write this concept in your Foldable. Be sure to include examples.

EXAMPLES Subtract Negative Integers

3 Find 12 – (–6).

$12 - (-6) = 12 +$ ______ To subtract –6, add ______.

$=$ ______ Simplify.

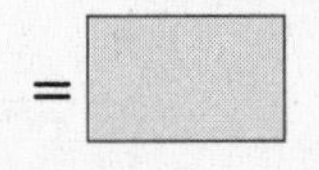

4 Find –21 – (–8).

$-21 - (-8) = -21 + 8$ To subtract ______, add ______.

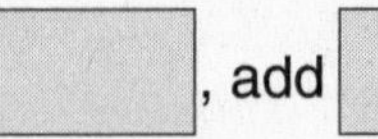

$= -13$ Simplify.

Check Your Progress **Subtract.**

a. $9 - (-4)$ **b.** $17 - (-6)$

EXAMPLE Evaluate an Expression

5 ALGEBRA **Evaluate $g - h$ if $g = -2$ and $h = -7$.**

$g - h =$ ____ $-$ ____ Replace ____ with -2 and h with ____.

$= -2 +$ ____ Subtract -7, add ____.

$=$ ____ Simplify.

Check Your Progress Evaluate $m - n$ if $m = -6$ and $n = 4$.

WRITE IT

Explain how you can use a number line to check the results of subtracting integers.

EXAMPLE

6 GEOGRAPHY **In Mongolia, the temperature can fall to $-45°C$ in January. The temperature in July may reach $40°C$. What is the difference between these two temperatures in Mongolia?**

To find the difference in temperatures, subtract the lower temperature from the higher temperature.

$40 - (-45) = 40$ ____ 45 To subtract -45,  ____ 45.

$=$ ____ Simplify.

So, the difference between the temperatures is ____.

Check Your Progress On a particular day in Anchorage, Alaska, the high temperature was $15°F$ and the low temperature was $-11°F$. What is the difference between these two temperatures for that day?

HOMEWORK ASSIGNMENT

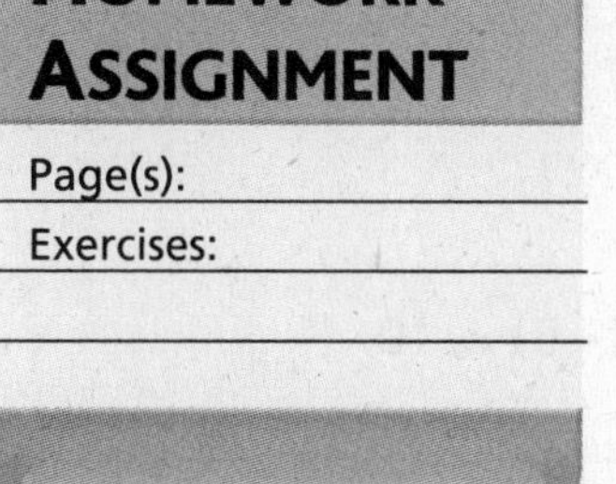

Page(s):

Exercises:

2-6 Multiplying Integers

MAIN IDEA

- Multiply integers.

EXAMPLES Multiply Integers with Different Signs

1 Find 5(–4).

$5(-4) =$ ______ The integers have ______ signs.

The product is ______.

2 Find –3(9).

$-3(9) =$ ______ The integers have ______ signs.

The product is ______.

Check Your Progress **Multiply.**

a. $3(-5)$ **b.** $-5(7)$

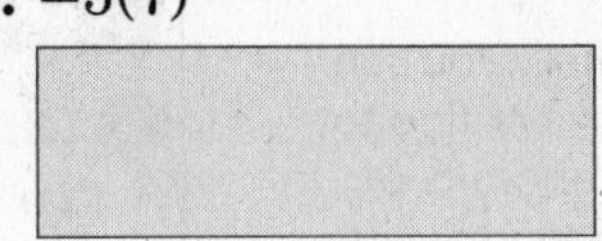

KEY CONCEPTS

Multiplying Integers with Different Signs The product of two integers with different signs is negative.

Multiply Integers with the Same Sign The product of two integers with the same sign is positive.

FOLDABLES Include these concepts on the Lesson 2–6 tab of your Foldable

EXAMPLES Multiply Integers with the Same Sign

3 Find –6(–8).

$-6(-8) =$ ______ The integers have the ______ sign.

The product is ______.

4 Find $(-8)^2$.

$(-8)^2 = (-8)$ ______ There are ______ factors of –8.

$=$ ______ The product is ______.

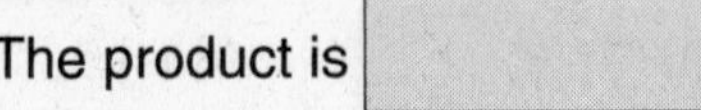

5 Find –2(–5)(–6).

$-2(-5)(-6) =$ ______ (-6) $-2(-5) = 10$

$=$ ______ $10(-6) = -60$

Check Your Progress **Multiply.**

a. $-4(-7)$ **b.** $(-5)^2$ **c.** $-7(-3)(-4)$

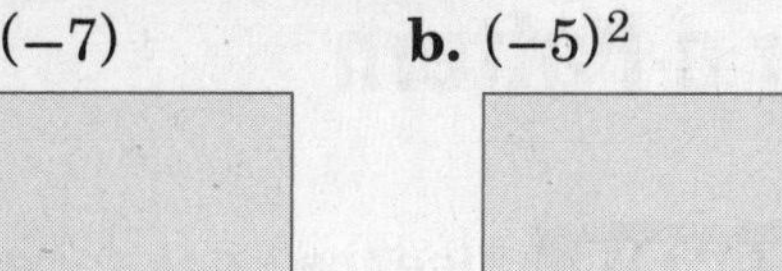

EXAMPLE

REMEMBER IT

When three variables are written without an operations sign, it means multiplication.

6 MINES A mine elevator descends at a rate of 300 feet per minute. How far below the earth's surface will the elevator be after 5 minutes?

If the elevator descends ______ feet per minute, then after 5 minutes, the elevator will be -300 (______) or $-1{,}500$ feet below the surface. Thus, the elevator will descend to ______ feet.

Check Your Progress **RETIREMENT** Mr. Rodriguez has \$78 deducted from his pay every month and placed in a savings account for his retirement. What integer represents a change in his savings account for these deductions after six months?

EXAMPLE **Evaluate Expressions**

7 ALGEBRA Evaluate abc if $a = -3$, $b = 5$, and $c = -8$.

$abc = (-3)(5)(-8)$ — Replace ______ with -3, b with ______, and c with ______.

$= (-15)(-8)$ — Multiply ______ and 5.

$=$ ______ — Multiply -15 and -8.

HOMEWORK ASSIGNMENT

Page(s):

Exercises:

Check Your Progress

Evaluate xyz if $x = -6$, $y = -2$, and $z = 4$.

2–7 Problem-Solving Investigation: Look for a Pattern

EXAMPLE Use the Look for a Pattern Strategy

MAIN IDEA

- Solve problems by looking for a pattern.

HAIR Lelani wants to grow an 11-inch ponytail. She has a 3-inch ponytail now, and her hair grows about one inch every two months. How long will it take for her ponytail to reach 11 inches?

UNDERSTAND You know the length of Lelani's ponytail now. You know how long Lelani wants her ponytail to grow and you know how fast her hair grows. You need to know how long it will take for her ponytail to reach ______ inches.

PLAN Look for a pattern. Then extend the pattern to find the solution.

SOLVE After the first two months, Lelani's ponytail will be 3 inches + ______ inch, or 4 inches. Every ______ months, her hair grows according to the pattern below.

3 in. 4 in. 5 in. 6 in. 7 in. 8 in. ______ 10 in. 11 in.

+1 +1 ______ +1 +1 +1 +1 +1

It will take eight sets of two months, or 16 months total, for Lelani's ponytail to reach ______ inches.

CHECK Lelani's ponytail grew from 3 inches to 11 inches, a difference of eight inches, in ______ months. Since one inch of growth requires two months and 8 × ______ = 16, the answer is correct.

Check Your Progress **RUNNING** Samuel ran 2 miles on his first day of training to run a marathon. On the third day, Samuel increased the length of his run by 1.5 miles. If this pattern continues every three days, how many miles will Samuel run on the 27th day?

______

HOMEWORK ASSIGNMENT

Page(s):

Exercises:

2–8 Dividing Integers

MAIN IDEA

- Divide integers.

KEY CONCEPTS

Dividing Integers with Different Signs The quotient of two integers with different signs is negative.

Dividing Integers with the Same Sign The quotient of two integers with the same sign is positive.

HOMEWORK ASSIGNMENT

Page(s):

Exercises:

EXAMPLES Dividing Integers with Different Signs

1 Find $51 \div (-3)$.

$51 \div (-3) =$ ______ The integers have ______ signs.

The ______ is negative.

2 Find $\frac{-121}{11}$.

$\frac{-121}{11} =$ ______ The ______ have different signs.

The quotient is ______.

EXAMPLE Dividing Integers with the Same Sign

3 Find $-12 \div (-2)$.

$-12 \div (-2) =$ ______ The integers have the ______ sign.

The quotient is ______.

Check Your Progress **Find each quotient.**

a. $36 \div (-9)$ **b.** $\frac{45}{-9}$ **c.** $-24 \div (-8)$

EXAMPLE

4 ALGEBRA Evaluate $-18 \div x$ if $x = -2$.

$-18 \div x = -18 \div ($ ______ $)$ Replace x with -2.

$=$ ______ Divide. The quotient is negative.

Check Your Progress **ALGEBRA** Evaluate $g \div h$ if $g = 21$ and $h = -3$.

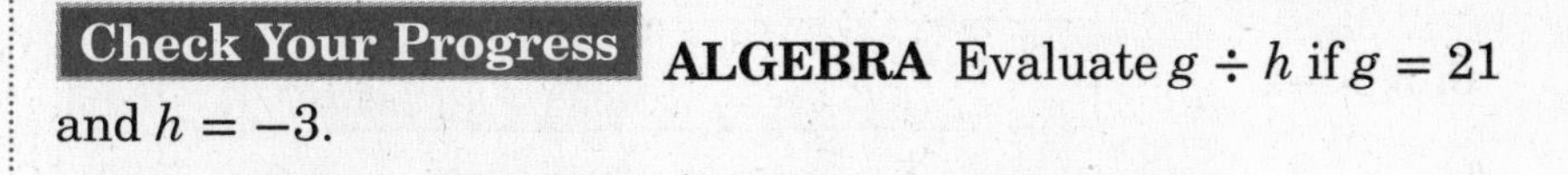

CHAPTER 2

BRINGING IT ALL TOGETHER

STUDY GUIDE

FOLDABLES	VOCABULARY PUZZLEMAKER	BUILD YOUR VOCABULARY
Use your **Chapter 2 Foldable** to help you study for your chapter test.	To make a crossword puzzle, word search, or jumble puzzle of the vocabulary words in Chapter 2, go to: glencoe.com	You can use your completed **Vocabulary Builder** (*pages 32–33*) to help you solve the puzzle.

2-1 Integers and Absolute Value

Express each of the following in words.

1. +7

2. −7

3. |7|

4. On the following number line, draw an oval around the *negative* integers and label them negative. Draw a rectangle around the *positive* integers and label them positive.

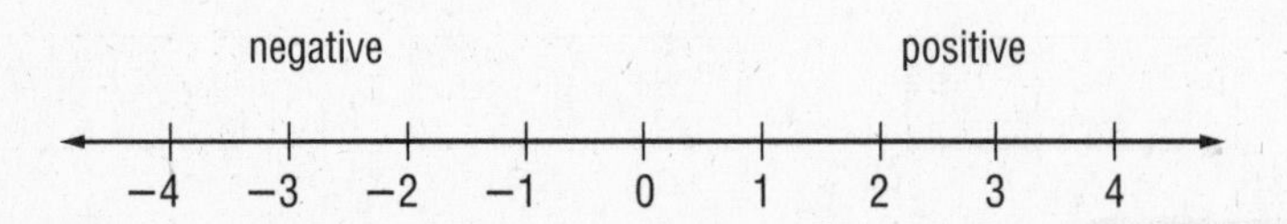

2-2 Comparing and Ordering Integers

Write each expression in words.

5. $-1 < 0$

6. $3 > -2$

2-3 The Coordinate Plane

Look at the coordinate plane at the right. Name the ordered pair for each point graphed.

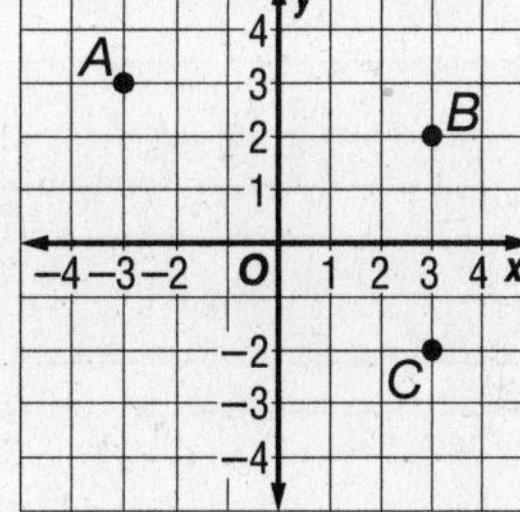

7. *A*

8. *B*

9. *C*

In the coordinate plane above, identify the quadrant in which each lies.

10. *A*

11. *B*

12. *C*

2-4 Adding Integers

Tell how you would solve each of the following on a number line, then add.

13. $-7 + (-9)$

14. $-7 + 9$

15. How many units away from 0 is the number 17?

16. How many units away from 0 is the number -17?

17. What are 17 and -17 called?

2-5 Subtracting Integers

Find each difference. Write an equivalent addition sentence for each.

18. $1 - 5$

19. $-2 - 1$

20. $-3 - 4$

2-6 Multiplying Integers

Choose the correct term to complete each sentence.

21. The product of two integers with different signs is (positive, negative).

22. The product of two integers with the same sign is (positive, negative).

Find each product.

23. $(-6)(-4)$ **24.** $-8(5)$ **25.** $-2(3)(-4)$

2-7 Problem-Solving Investigation: Look for a Pattern

26. CANS A display of soup cans at the end of a store aisle contains 1 can in the top row and 2 cans in each additional row beneath it. If there are 6 rows in the display, how many cans are in the sixth row?

2-8 Dividing Integers

Write two division sentences for each of the following multiplication sentences.

27. $6(-3) = 18$

28. $-21(-2) = 42$

ARE YOU READY FOR THE CHAPTER TEST?

Math Online

Visit **glencoe.com** to access your textbook, more examples, self-check quizzes, and practice tests to help you study the concepts in Chapter 2.

Check the one that applies. Suggestions to help you study are given with each item.

☐ **I completed the review of all or most lessons without using my notes or asking for help.**

- You are probably ready for the Chapter Test.
- You may want to take the Chapter 2 Practice Test on page 123 of your textbook as a final check.

☐ **I used my Foldables or Study Notebook to complete the review of all or most lessons.**

- You should complete the Chapter 2 Study Guide and Review on pages 119–122 of your textbook.
- If you are unsure of any concepts or skills, refer back to the specific lesson(s).
- You may want to take the Chapter 2 Practice Test on page 123 of your textbook.

☐ **I asked for help from someone else to complete the review of all or most lessons.**

- You should review the examples and concepts in your Study Notebook and Chapter 2 Foldables.
- Then complete the Chapter 2 Study Guide and Review on pages 119–122 of your textbook.
- If you are unsure of any concepts or skills, refer back to the specific lesson(s).
- You may also want to take the Chapter 2 Practice Test on page 123 of your textbook.

Student Signature

Parent/Guardian Signature

Teacher Signature

Algebra: Linear Equations and Functions

Use the instructions below to make a Foldable to help you organize your notes as you study the chapter. You will see Foldable reminders in the margin of this Interactive Study Notebook to help you in taking notes.

Begin with a sheet of 11" × 17" paper.

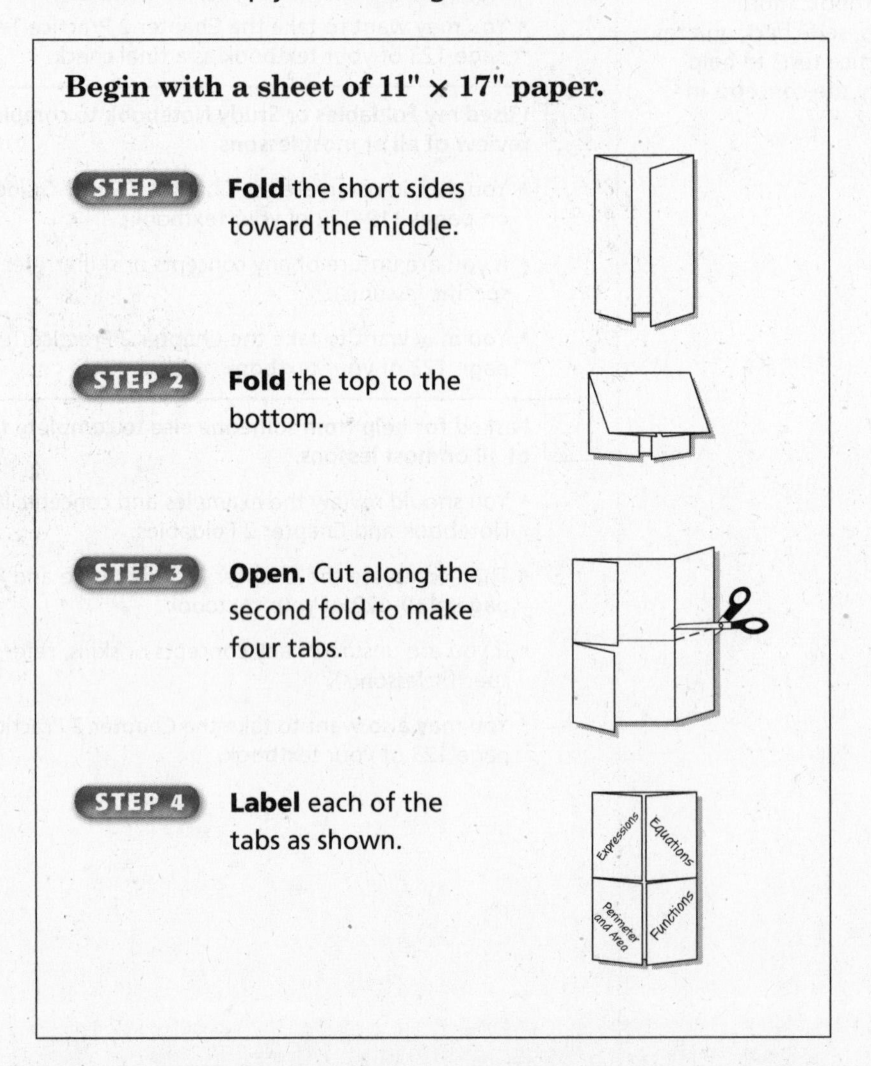

Fold the short sides toward the middle.

Fold the top to the bottom.

Open. Cut along the second fold to make four tabs.

Label each of the tabs as shown.

NOTE-TAKING TIP: When you take notes, listen or read for main ideas. Then record those ideas in a simplified form for future reference.

BUILD YOUR VOCABULARY

This is an alphabetical list of new vocabulary terms you will learn in Chapter 3. As you complete the study notes for the chapter, you will see Build Your Vocabulary reminders to complete each term's definition or description on these pages. Remember to add the textbook page number in the second column for reference when you study.

Vocabulary Term	Found on Page	Definition	Description or Example
Addition Property of Equality			
Division Property of Equality			
formula			
linear equation			

(continued on the next page)

Vocabulary Term	Found on Page	Definition	Description or Example
Subtraction Property of Equality			
two-step equation			
work backward strategy			

3–1 Writing Expressions and Equations

MAIN IDEA

- Write verbal phrases and sentences as simple algebraic expressions and equations.

FOLDABLES

ORGANIZE IT

Write two phrases and their algebraic expressions under the **Expressions** tab.

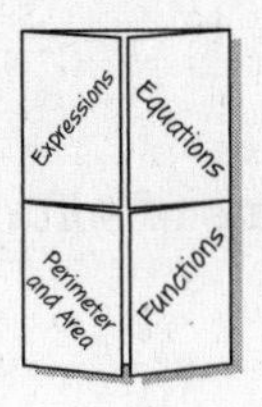

EXAMPLE Write a Phrase as an Expression

1 Write the phrase *twenty dollars less the price of a movie ticket* as an algebraic expression.

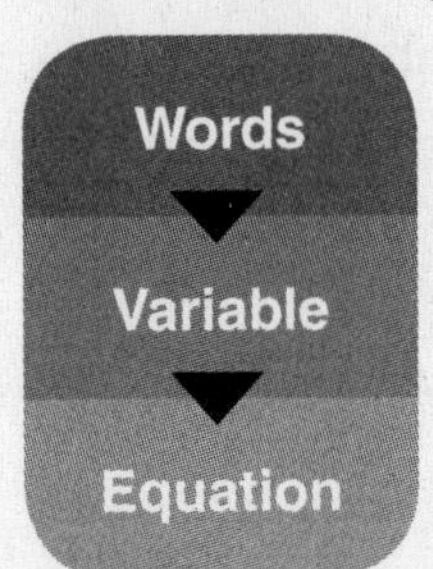

twenty dollars less the price of a movie ticket

Let ______ = the price of a movie ticket.

______

Check Your Progress Write the phrase *five more inches of snow than last year's snowfall* as an algebraic expression.

______

EXAMPLES Write Sentences as Equations

Write each sentence as an algebraic equation.

2 A number less 4 is 12.

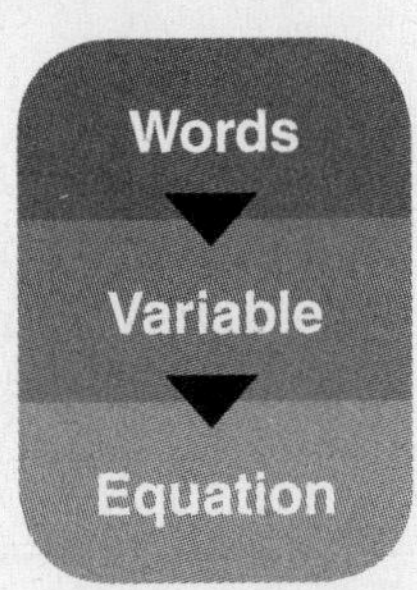

A number less 4 is 12.

Let ______ represent a number.

______

3 Twice a number is 18.

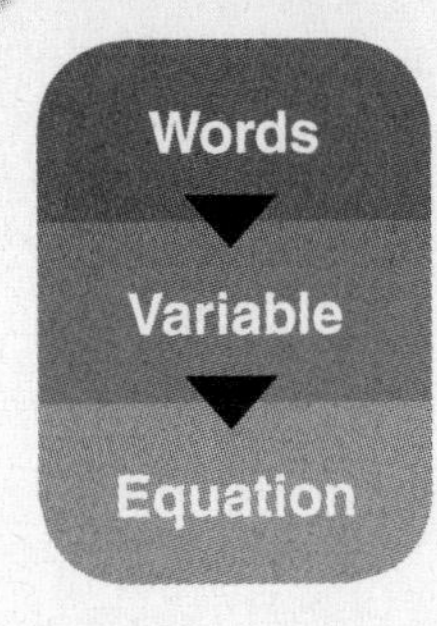

Twice a number is 18.

Let ______ represent a number.

______

Check Your Progress **Write each sentence as an algebraic equation.**

a. Eight less than a number is 12.

b. Four times a number equals 96.

EXAMPLE

4 FOOD An average American adult drinks more soft drinks than any other beverage each year. Three times the number of gallons of soft drinks plus 27 is equal to the total 183 gallons of beverages consumed. Write the equation that models this situation.

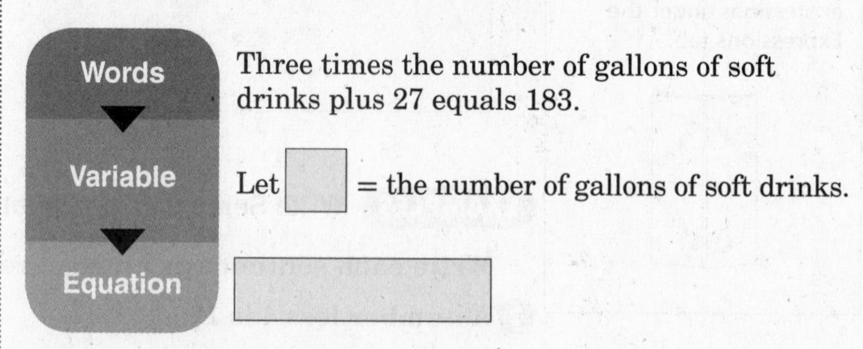

Check Your Progress **EXERCISE** It is estimated that American adults spend an average of 8 hours per month exercising. This is 26 hours less than twice the number of hours spent watching television each month. Write an equation that models this situation.

HOMEWORK ASSIGNMENT

Page(s):

Exercises:

3–2

Solving Addition and Subtraction Equations

MAIN IDEA

- Solve addition and subtraction equations.

KEY CONCEPTS

Subtraction Property of Equality If you subtract the same number from each side of an equation, the two sides remain equal.

Addition Property of Equality If you add the same number to each side of an equation, the two sides remain equal.

FOLDABLES Write these properties in your own words under the **Equations** tab.

EXAMPLES Solve an Addition Equation

1 **Solve $14 + y = 20$. Check your solution.**

$14 \quad + \quad y \quad = \quad 20$ — Write the equation.

____ ____ — ____ 14 from each side. Simplify.

____ = ____

Check

$14 + y = 20$ — Write the original equation.

$14 +$ ____ $\stackrel{?}{=} 20$ — Replace *y* with ____.

____ $= 20$ ✓ — Simplify.

The solution is ____.

2 **Solve $a + 7 = 6$. Check your solution.**

$a \quad + \quad 7 \quad = \quad 6$ — Write the equation.

____ ____ — Subtract ____ from each side.

____ = ____ — Simplify.

Check

$a + 7 = 6$ — Write the original equation.

____ $+ 7 \stackrel{?}{=} 6$ — Replace *a* with ____.

____ $= 6$ ✓ — Simplify. The solution is .

Check Your Progress **Solve each equation.**

a. $-6 = x + 4$

b. $m + 9 = 22$

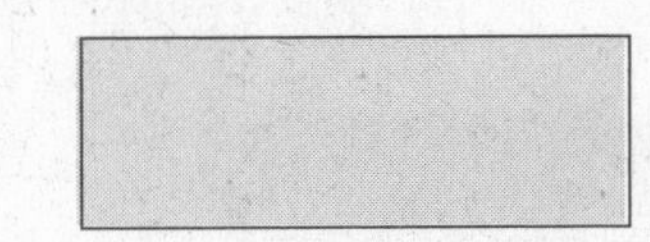

EXAMPLE

3 FRUIT **A grapefruit weighs 11 ounces, which is 6 ounces more than an apple. How much does the apple weigh?**

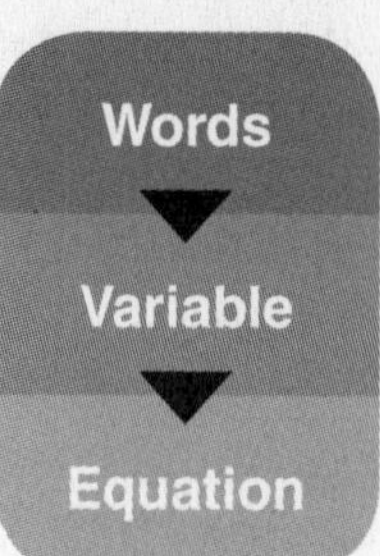

A grapefruit's weight	is	ounces	more than	an apple's weight.

Let a represent the apple's weight.

11	= 6	+	a

Write the equation.

$-6 \quad -6$ Subtract from each side.

$5 = a$ Simplify.

The apple weighs ounces.

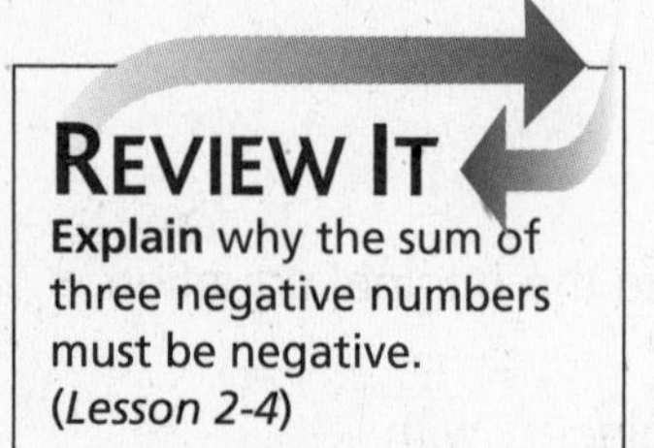

REVIEW IT

Explain why the sum of three negative numbers must be negative. (*Lesson 2-4*)

Check Your Progress **EXERCISE** Cedric ran 17 miles this week, which is 9 more miles than he ran last week. How many miles did he run last week?

EXAMPLE **Solve a Subtraction Equation**

4 **Solve $12 = z - 8$.**

$12 = z - 8$ Write the equation.

$+8 \quad +8$ Add 8 to each side.

$= z$ Simplify.

The solution is .

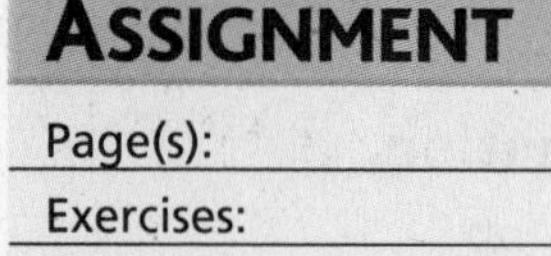

HOMEWORK ASSIGNMENT

Page(s):

Exercises:

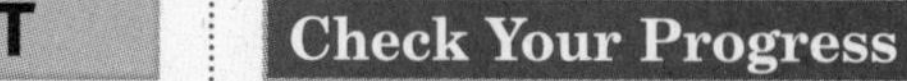

Check Your Progress Solve $w - 5 = 27$.

3-3 Solving Multiplication Equations

EXAMPLES Solving Multiplication Equations

MAIN IDEA

- Solve multiplication equations.

1 Solve $39 = 3y$. Check your solution.

$39 = 3y$ Write the equation.

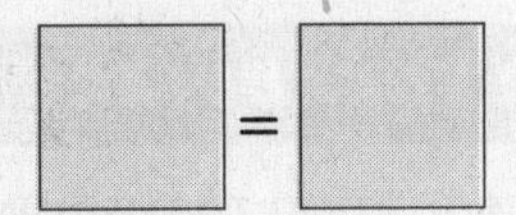

Divide each side of the equation by ____.

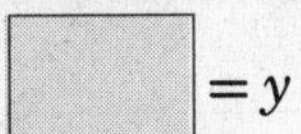

____ $\div 3 =$ ____

Check

$39 = 3y$ Write the equation.

$39 \stackrel{?}{=} 3$ ____ Replace y with ____. Is this sentence true?

$39 =$ ____ ✓

So, the solution is ____.

KEY CONCEPT

Division Property of Equality If you divide each side of an equation by the same nonzero number, the two sides remain equal.

FOLDABLES Record the Division Property of Equality in your own words under the Equation tab.

2 Solve $-4z = 60$. Check your solution.

$-4z = 60$ Write the equation.

____ = ____ Divide each side of the equation by .

$z =$ ____ $60 \div (-4) =$ ____

Check

$-4z = 60$ Write the equation.

$-4($ ____ $) \stackrel{?}{=} 60$  Replace z with ____. Is this sentence true?

____ $= 60$ ✓

So, the solution is ____.

Check Your Progress **Solve each equation. Check your solution.**

a. $6m = 42$

b. $-64 = -16b$

BUILD YOUR VOCABULARY (pages 55–56)

A **formula** is an equation that shows the relationship among certain quantities.

EXAMPLE

3 SWIMMING **Ms. Wang swims at a speed of 0.6 mph. At this rate, how long will it take her to swim 3 miles?**

You are asked to find the time t it will take to swim a distance d of 3 miles at a rate r of 0.6 mph.

$d = rt$	Write the equation.
$3 = 0.6t$	Replace d with ______ and r with ______.
$\frac{3}{0.6} = \frac{0.6t}{0.6}$	Divide each side by 0.6.
______ $= t$	$3 \div 0.6 = 5$

It would take Ms. Wang ______ hours to swim 3 miles.

Check Your Progress **COOKIES** Debbie spends \$6.85 on cookies at the bakery. The cookies are priced at \$2.74 per pound. How many pounds of cookies did Debbie buy?

HOMEWORK ASSIGNMENT

Page(s):

Exercises:

3-4 Problem-Solving Investigation: Work Backward

EXAMPLE Use the Work Backward Strategy

MAIN IDEA

- Solve problems using the work backward strategy.

SHOPPING **Lucy and Elena went to the mall. Each girl bought a CD for $16.50, a popcorn for $3.50, and a drink for $2.50. Altogether, they had $5.00 left over. How much money did they take to the mall?**

UNDERSTAND You know that they had ______ left over and how much they spent on each item. You need to know how much they took to the mall.

PLAN Start with the end result and work backward.

SOLVE They had $5.00 left.

Undo the two drinks for $2.50 each. $\$5 + 2(\$2.50) =$ ______

Undo the two popcorns for ______ each. $\$10 + 2(\$3.50) =$ ______

Undo the two CDs for $16.50 each. $\$17 + 2(\$16.50) =$ ______

So, they took ______ to the mall.

CHECK Assume they started with $50. After buying two CDs, they had $\$50 - 2(\$16.50)$ or ______.

After buying two popcorns, they had $\$17 - 2(\quad)$ or $10. After buying two drinks, they had $\$10 - 2(\$2.50)$ or $5. So, the answer is correct.

Check Your Progress **AIRPORT** Jack needs to go home from work to pack before heading to the airport. He wants to be at the airport by 1:15 P.M. It takes him 20 minutes to drive home from work, 30 minutes to pack, and 45 minutes to get to the airport from home. What time should he leave work?

HOMEWORK ASSIGNMENT

Page(s):

Exercises:

3–5 Solving Two-Step Equations

BUILD YOUR VOCABULARY (pages 55–56)

A **two-step equation** has ______ different ______.

MAIN IDEA

- Solve two-step equations.

EXAMPLES Solve Two-Step Equations

1 Solve $4x + 3 = 19$. Check your solution.

$4x \quad + \quad 3 \quad = \quad 19$ — Write the equation.

______ ______ — Subtract ______ from each side.

______ = ______ — Simplify.

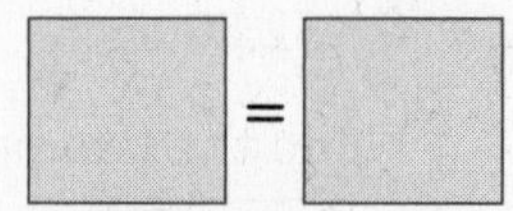

______ = ______ — Divide each side by ______.

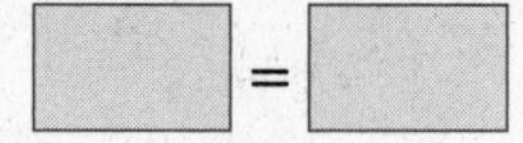

______ = ______ — Simplify.

Check

$4x + 3 = 19$ — Write the original equation.

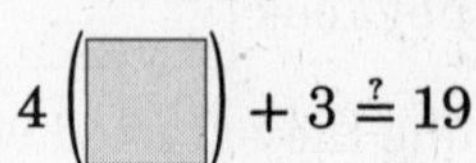

$4(\ \_\_\_\_\ ) + 3 \stackrel{?}{=} 19$ — Replace x with ______.

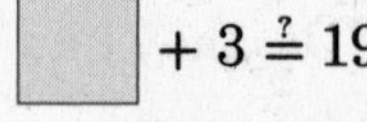

______ $+ 3 \stackrel{?}{=} 19$ — Simplify.

______ $= 19$ ✓

The solution is ______.

WRITE IT

What is the name of the property that allows you to subtract the same number from each side of an equation?

______

______

______

2 Solve $6 + 5y = 26$.

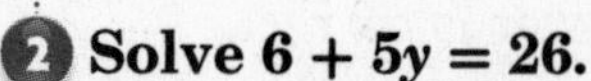

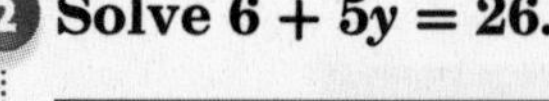

______ — Write the equation.

$\underline{-6 \qquad\quad -6}$ — Subtract ______ from each side.

$5y = 20$ — Simplify.

$\frac{5y}{5} = \frac{20}{5}$ — Divide each side by ______.

$y =$ ______ — Simplify.

3 **Solve $-3c + 9 = 3$.**

$-3c \quad + \quad 9 \quad = \quad 3$ Write the equation.

Subtract from each side.

= Simplify.

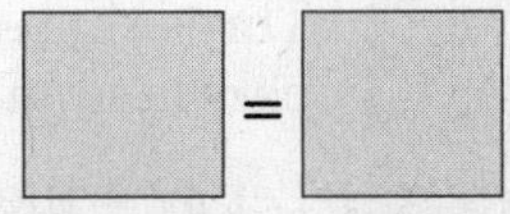

= Divide each side by .

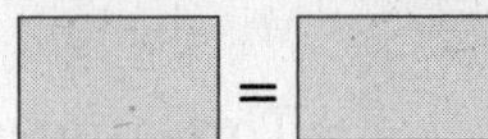

= Simplify.

The solution is .

4 **Solve $0 = 6 + 3t$.**

$0 = 6 + 3t$ Write the equation.

$-6 \quad -6$

= Simplify.

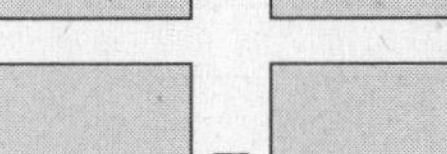

= each side by .

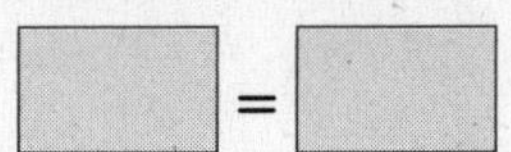

= Simplify.

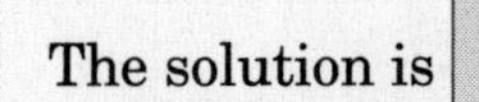

The solution is .

REMEMBER IT

Always check your solutions by replacing the variable with your answer and simplifying.

Check Your Progress **Solve each equation.**

a. $3t - 7 = 14$

b. $4 + 2w = 18$

c. $-8k + 7 = 31$

d. $0 = -4x + 32$

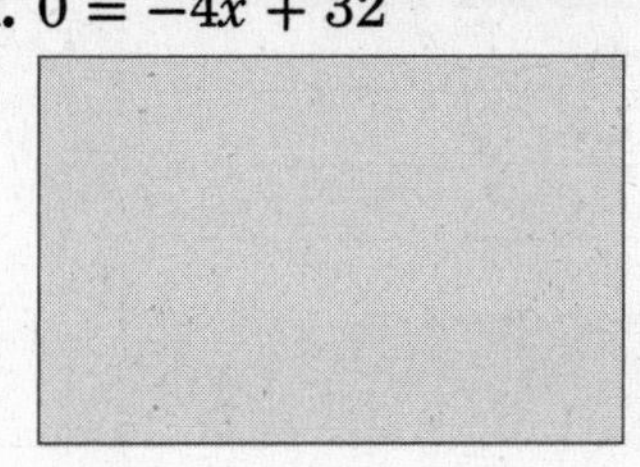

EXAMPLE

5 PARKS There are 76 thousand acres of state parkland in Georgia. This is 4 thousand acres more than three times the number of acres of state parkland in Mississippi. How many acres of state parkland are there in Mississippi?

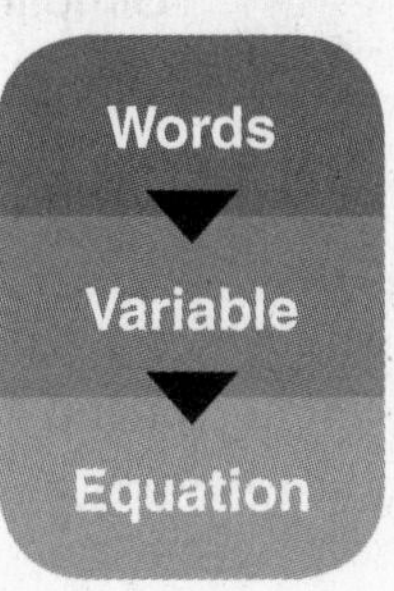

Three times the number of acres of state parkland in Mississippi plus 4,000 is 76,000.

Let m = the acres of state parkland in Mississippi.

Three times the number of acres of parkland in Mississippi	plus	4,000	is	76,000.
___	___	4,000	=	76,000

___ + 4,000 = 76,000 Write the equation.

___ ___ Subtract ___ from each side.

___ = ___ Simplify.

___ = ___ Divide each side by ___.

___ = ___ Simplify.

There are ___ acres of state parkland in Mississippi.

Check Your Progress **BASEBALL** Matthew had 64 hits during last year's baseball season. This was 8 less than twice the number of hits Gregory had. How many hits did Gregory have during last year's baseball season?

HOMEWORK ASSIGNMENT

Page(s):

Exercises:

3–6 Measurement: Perimeter and Area

MAIN IDEA

- Find the perimeters and areas of figures.

BUILD YOUR VOCABULARY (pages 55–56)

The ______ around a geometric figure is called the **perimeter**.

EXAMPLE Find the Perimeter of a Rectangle

1 Find the perimeter of the rectangle.

2 ft
18 ft

$P = 2\ell + 2w$ Perimeter of a rectangle

$P = 2(18) + 2(2)$ $\ell =$ ______, $w =$ ______

$P =$ ______ $+$ ______ Multiply.

$P =$ ______ Add.

The perimeter is 40 ______.

KEY CONCEPT

Perimeter of a Rectangle The perimeter P of a rectangle is twice the sum of the length ℓ and width w.

Check Your Progress Find the perimeter of a rectangle with a length of 2.35 centimeters and a width of 11.9 centimeters.

EXAMPLE

2 ART A painting has a perimeter of 68 inches. If the width of the painting is 13 inches, what is its length?

$P = 2\ell + 2w$ Perimeter of a rectangle

$68 = 2\ell + 2($ ______ $)$ Replace P with 68 and w with 13.

$68 = 2\ell +$ ______ Multiply.

(continued on the next page)

$68 - 26 = 2\ell + 26 - 26$	Subtract 26 from each side.
$\qquad = 2\ell$	Simplify.
$21 = \ell$	Divide each side by 2.

Check Your Progress **GARDENS** A tomato garden has a perimeter of 22.2 feet. If the length of the garden is 6.3 feet, find the width.

BUILD YOUR VOCABULARY (pages 55–56)

The **area** is the measure of the ________ enclosed by a figure.

EXAMPLE Find The Area of a Rectangle

KEY CONCEPT

Area of a Rectangle The area *A* of a rectangle is the product of the length ℓ and width w.

3 FRESHWATER **Find the area of the surface of the reservoir shown below.**

0.625 mi

4 mi

$A = \ell \cdot w$	Area of a ________
$A = \_\_\_ \cdot \_\_\_$	Replace ℓ with 4 and w with ________.
$A = \_\_\_$	________.

The area is 2.5 ________.

HOMEWORK ASSIGNMENT

Page(s):

Exercises:

Check Your Progress

PAINTING Sue is painting a wall that measures 18.25 feet long and 8 feet high. Find the area of the surface Sue will be painting.

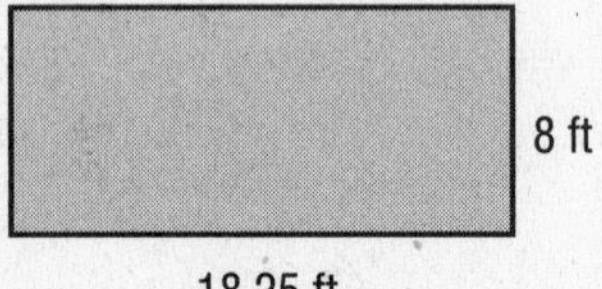

3-7 Functions and Graphs

MAIN IDEA

- Graph linear equations.

REMEMBER IT

When x and y are used in an equation, x usually represents the input and y usually represents the output.

EXAMPLE

1 WORK **The table shows the number of hours Abby worked and her corresponding earnings. Make a graph of the data to show the relationship between the number of hours Abby worked and her earnings.**

The ordered pairs (1, 6), (____, 12), (3, ____), and (4, 24) represent the function. Graph the ordered pairs.

Number of Hours	Earnings ($)
1	6
2	12
3	18
4	24

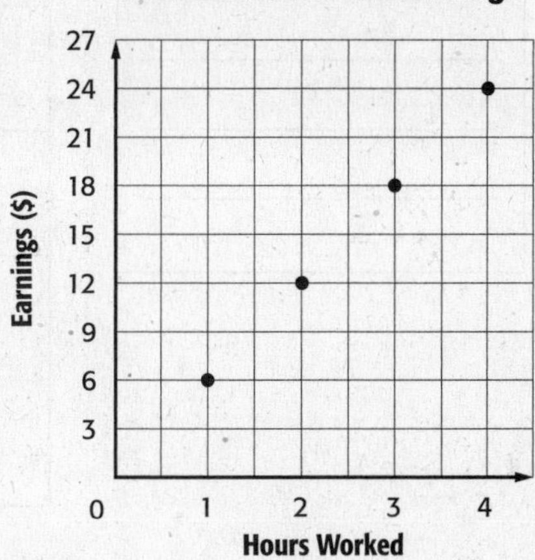

Check Your Progress **VIDEOS** Make a graph of the data in the table that shows the relationship between the amount David would pay and the number of movies he rents.

Number of Videos	Amount ($)
1	$3.50
2	$7.00
3	$10.50
4	$14.00

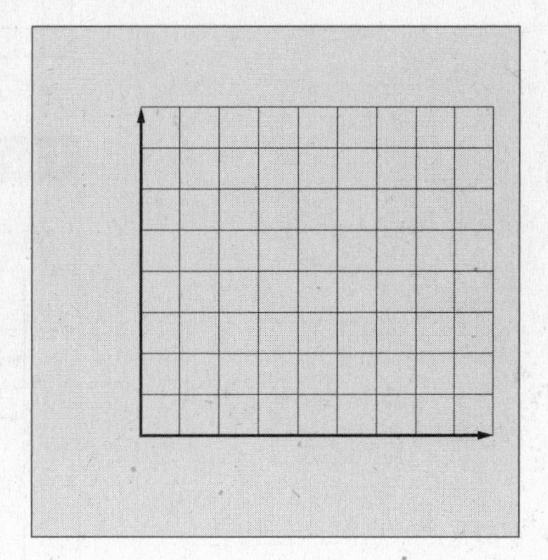

BUILD YOUR VOCABULARY (pages 55–56)

An equation like $y = 2x + 1$ is a **linear equation** because

the ________ is a ________ line.

WRITE IT

How many points are needed to graph a line? Why is it a good idea to graph more?

____

____

____

____

EXAMPLE Graph Solutions of Linear Equations

2 **Graph $y = x + 3$.**

Select any four values for the input x. We chose 2, 1, 0, and −1. Substitute these values for x to find the output y.

x	$x + 3$	y	(x, y)
2	____ + 3	____	(2, 5)
1	____ + 3	4	____
0	0 + 3	____	____
−1	____ + 3	2	____

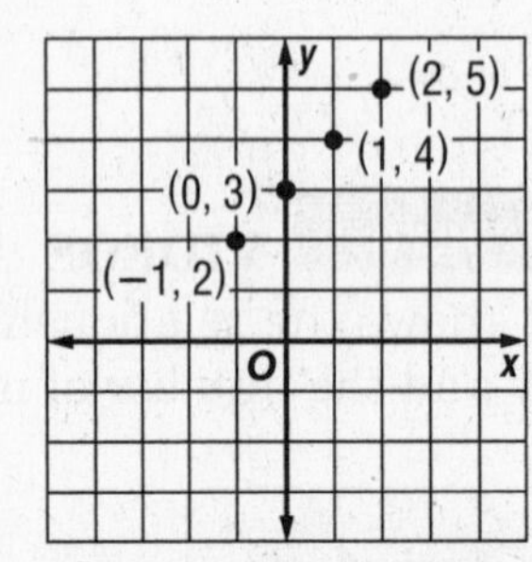

Four solutions are (2, 5), ____, ____, and ____.

Check Your Progress Graph $y = 3x - 2$.

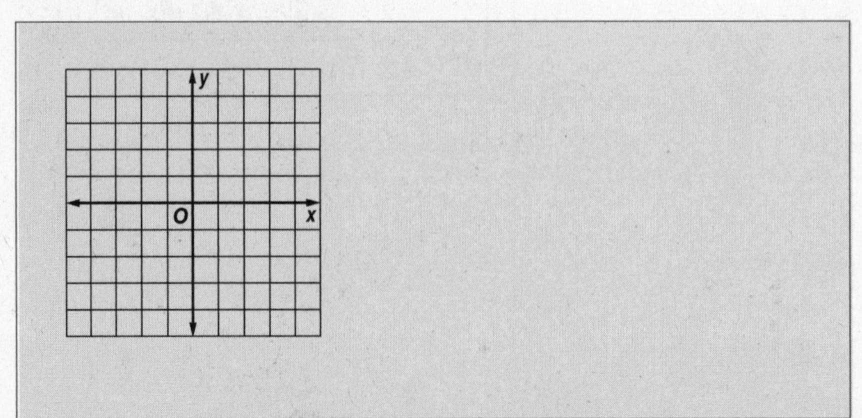

EXAMPLE Represent Real-World Functions

3 **ANIMALS Blue whales can reach a speed of 30 miles per hour. The equation $d = 30t$ describes the distance d that a whale swimming at that speed can travel in time t. Assuming that a whale can maintain that speed, represent the function with a graph.**

Step 1 Select four values for t. Select only positive numbers since t represents time. Make a function table.

t	$30t$	d	(t, d)
2	30(2)		(2, 60)
3	30(3)	90	
5	30(5)		
6	30	180	

Step 2 Graph the ordered pairs and draw a line through the points.

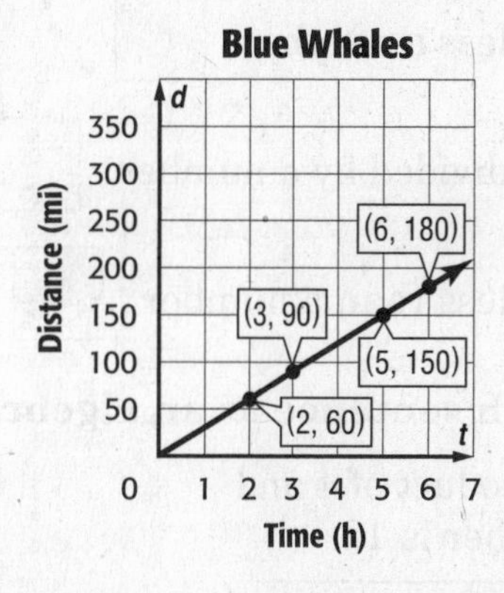

Check Your Progress **TRAVEL** Susie takes a car trip traveling at an average speed of 55 miles per hour. The equation $d = 55t$ describes the distance d that Susie travels in time t. Represent this function with a graph.

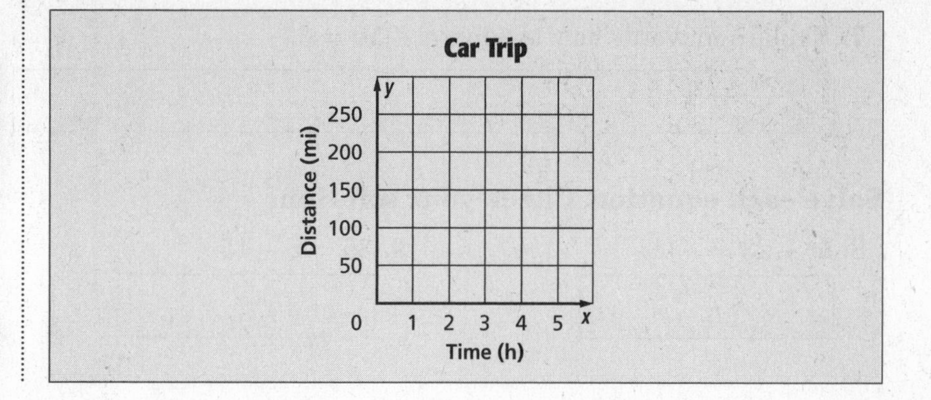

HOMEWORK ASSIGNMENT

Page(s):

Exercises:

CHAPTER 3 BRINGING IT ALL TOGETHER

STUDY GUIDE

FOLDABLES	VOCABULARY PUZZLEMAKER	BUILD YOUR VOCABULARY
Use your **Chapter 3 Foldable** to help you study for your chapter test.	To make a crossword puzzle, word search, or jumble puzzle of the vocabulary words in Chapter 3, go to: glencoe.com	You can use your completed **Vocabulary Builder** (*pages 55–56*) to help you solve the puzzle.

3-1 Writing Expressions and Equations

Match the phrases with the algebraic expressions that represent them.

1. seven plus a number

2. seven less a number

3. seven divided by a number

4. seven less than a number

a. $7 - n$

b. $7 \cdot n$

c. $n - 7$

d. $\frac{n}{7}$

e. $7 + n$

Write each sentence as an algebraic equation.

5. The product of 4 and a number is 12.

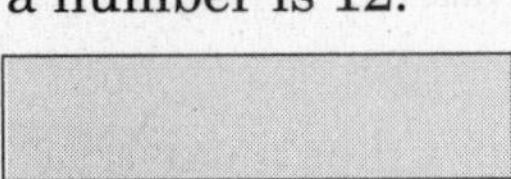

6. Twenty divided by y is equal to -10.

3-2 Solving Addition and Subtraction Equations

7. Explain in words how to solve $a - 10 = 3$.

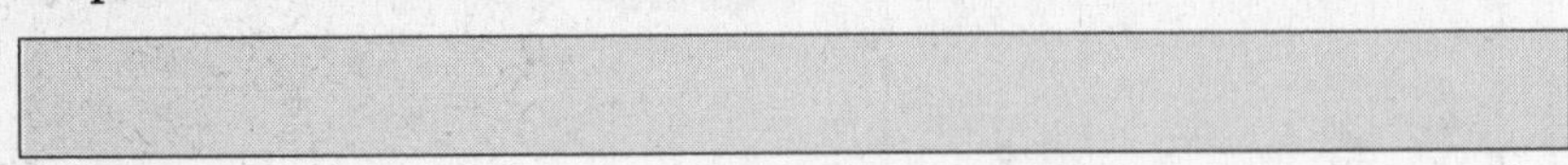

Solve each equation. Check your solution.

8. $w + 23 = -11$

9. $35 = z - 15$

3-3 Solving Multiplication Equations

10. To solve $-27 = -3d$, divide each side by ____.

Solve each equation. Check your solution.

11. $36 = 6k$

12. $-7z = 28$

3-4 Problem-Solving Investigation: Work Backward

13. **AGE** Bradley is four years older than his brother Philip. Philip is 7 years younger than Kailey, who is 2 years older than Taneesha. If Taneesha is 11 years old, how old is Bradley?

3-5 Solving Two-Step Equations

14. Describe in words each step shown for solving $12 + 7s = -9$.

$$12 + 7s = -9$$

$$\underline{-12 \qquad -12}$$

$$7s = -21$$

$$\frac{7s}{7} = \frac{-21}{7}$$

$$s = -3$$

15. Number the steps in the correct order for solving the equation $-4v + 11 = -5$.

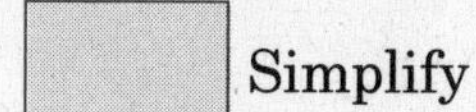

Simplify

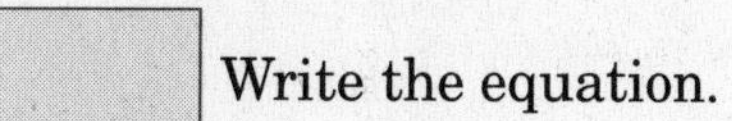

Write the equation.

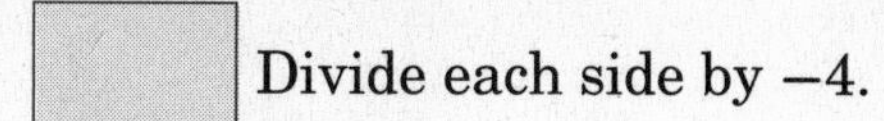

Divide each side by -4.

Simplify.

Subtract 11 from each side.

Check the solution.

3-6

Measurement: Perimeter and Area

Find the perimeter and area of each rectangle.

16.

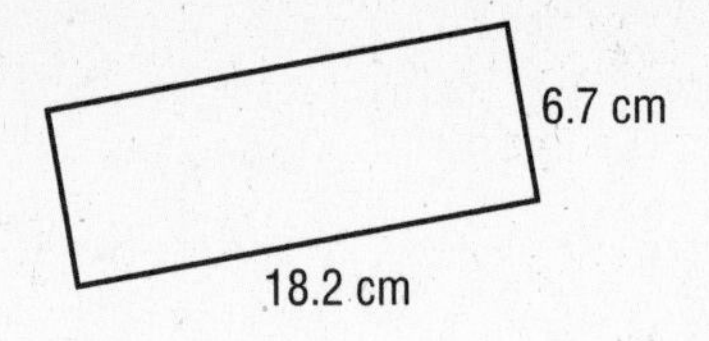

17.

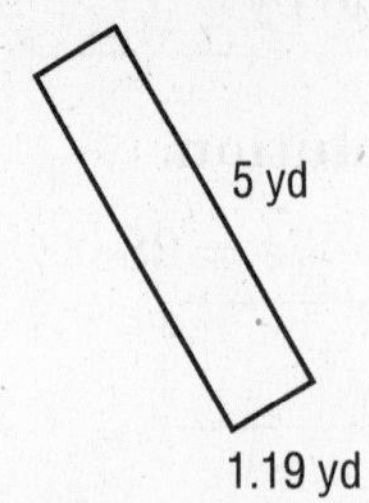

18. **FRAMING** Marcia wants to frame her favorite painting. If the frame is 3.25 feet wide and the perimeter is 15.7 feet, find the width of the frame.

3-7

Functions and Graphs

19. Complete the function table. Then graph the function.

x	$2x - 1$	y
−1		
0		
1		

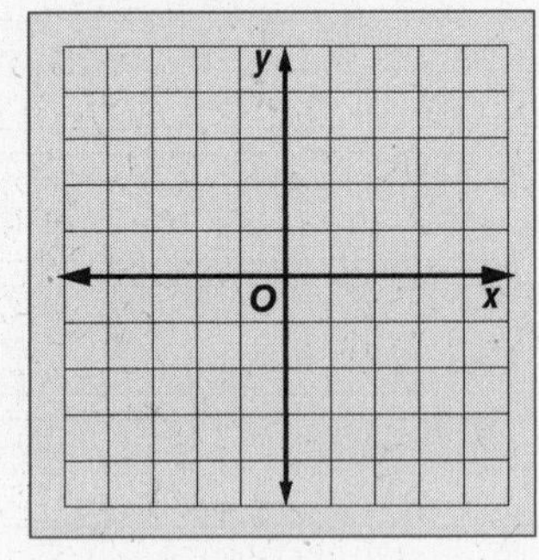

ARE YOU READY FOR THE CHAPTER TEST?

Math Online

Visit **glencoe.com** to access your textbook, more examples, self-check quizzes, and practice tests to help you study the concepts in Chapter 3.

Check the one that applies. Suggestions to help you study are given with each item.

☐ **I completed the review of all or most lessons without using my notes or asking for help.**

- You are probably ready for the Chapter Test.
- You may want to take the Chapter 3 Practice Test on page 173 of your textbook as a final check.

☐ **I used my Foldables or Study Notebook to complete the review of all or most lessons.**

- You should complete the Chapter 3 Study Guide and Review on pages 169–172 of your textbook.
- If you are unsure of any concepts or skills, refer back to the specific lesson(s).
- You may want to take the Chapter 3 Practice Test on page 173.

☐ **I asked for help from someone else to complete the review of all or most lessons.**

- You should review the examples and concepts in your Study Notebook and Chapter 3 Foldable.
- Then complete the Chapter 3 Study Guide and Review on pages 169–172 of your textbook.
- If you are unsure of any concepts or skills, refer back to the specific lesson(s).
- You may also want to take the Chapter 3 Practice Test on page 173.

Student Signature

Parent/Guardian Signature

Teacher Signature

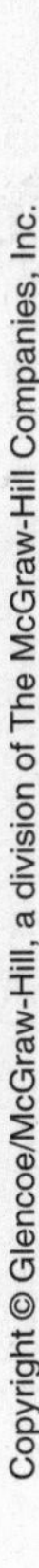

Fractions, Decimals, and Percents

Use the instructions below to make a Foldable to help you organize your notes as you study the chapter. You will see Foldable reminders in the margin this Interactive Study Notebook to help you in taking notes.

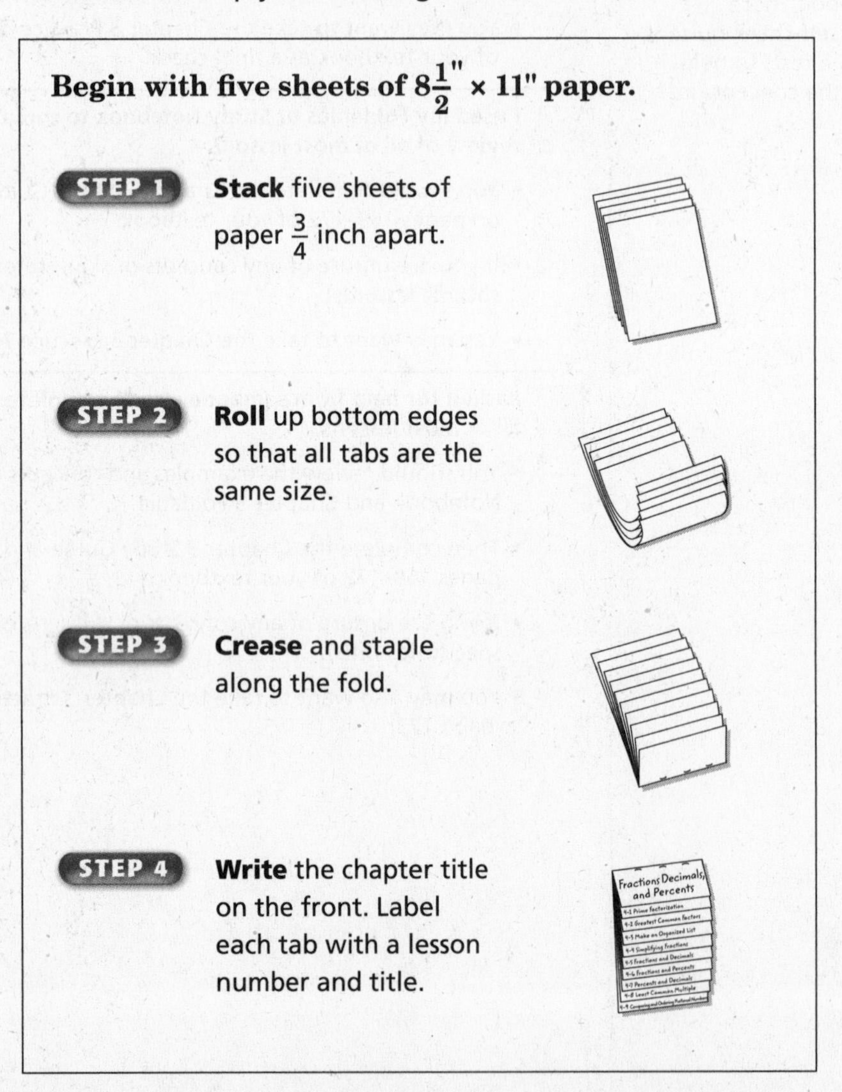

Begin with five sheets of $8\frac{1}{2}" \times 11"$ paper.

STEP 1 **Stack** five sheets of paper $\frac{3}{4}$ inch apart.

STEP 2 **Roll** up bottom edges so that all tabs are the same size.

STEP 3 **Crease** and staple along the fold.

STEP 4 **Write** the chapter title on the front. Label each tab with a lesson number and title.

NOTE-TAKING TIP: Before each lesson, skim through the lesson and write any questions that come to mind in your notes. As you work through the lesson, record the answer to your question.

CHAPTER 4

BUILD YOUR VOCABULARY

This is an alphabetical list of new vocabulary terms you will learn in Chapter 4. As you complete the study notes for the chapter, you will see Build Your Vocabulary reminders to complete each term's definition or description on these pages. Remember to add the textbook page number in the second column for reference when you study.

Vocabulary Term	Found on Page	Definition	Description or Example
bar notation			
common denominator			
composite number [kahm-PAH-zuht]			
equivalent [ih-KWIH-vuh-luhnt] fractions			
factor tree			
greatest common factor (GCF)			
least common denominator (LCD)			
least common multiple (LCM)			
multiple			

(continued on the next page)

Vocabulary Term	Found on Page	Definition	Description or Example
percent			
prime factorization			
prime number			
ratio			
rational number			
repeating decimal			
simplest form			
terminating decimal			

4-1 Prime Factorization

MAIN IDEA

- Find the prime factorization of a composite number.

BUILD YOUR VOCABULARY (pages 77–78)

A **prime number** is a whole number greater than 1 that has exactly ______ factors, ______ and ______.

A **composite number** is a whole number greater than ______ that has more than ______ factors.

Every ______ number can be written as a product of prime numbers exactly one way called the **prime factorization.**

A **factor tree** can be used to find the factorization.

FOLDABLES

ORGANIZE IT

Under the tab for Lesson 4-1, give examples of prime and composite numbers. Be sure to explain how to tell a prime number from a composite number.

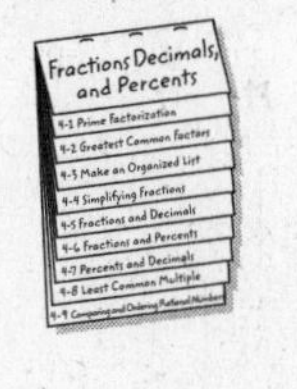

EXAMPLES **Identify Numbers as Prime or Composite**

Determine whether each number is *prime* or *composite*.

1 **63**

63 has six factors: 1, ______, 7, ______, 21, and ______.

So, it is ______.

2 **29**

29 has only two factors: ______ and ______.

So, it is ______.

Check Your Progress **Determine whether each number is *prime* or *composite*.**

a. 41

b. 24

EXAMPLE Find the Prime Factorization

REMEMBER IT

Multiplication is commutative, so the order of factors does not matter.

3 **Find the prime factorization of 100.**

To find the prime factorization, you can use a factor tree or divide by prime numbers. Let's use a factor tree.

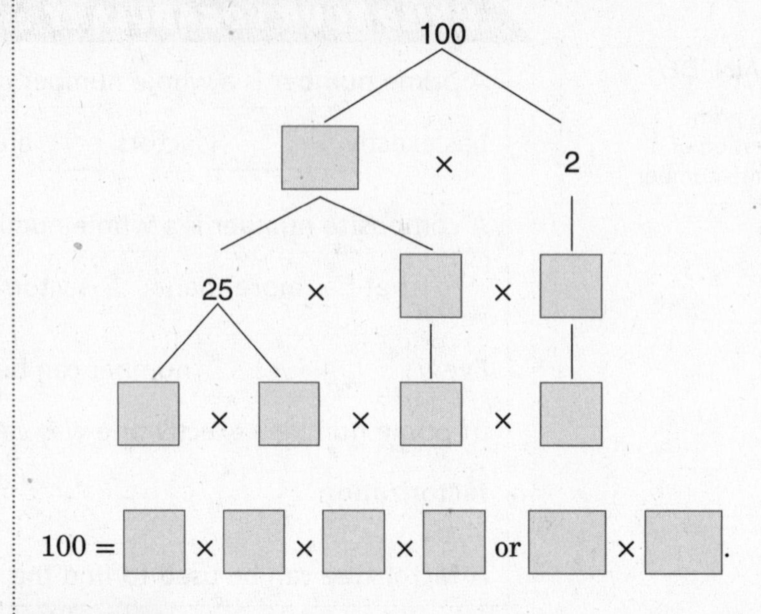

100 = ☐ × ☐ × ☐ × ☐ or ☐ × ☐.

EXAMPLE Find an Algebraic Expression

4 ALGEBRA **Factor $21m^2n$.**

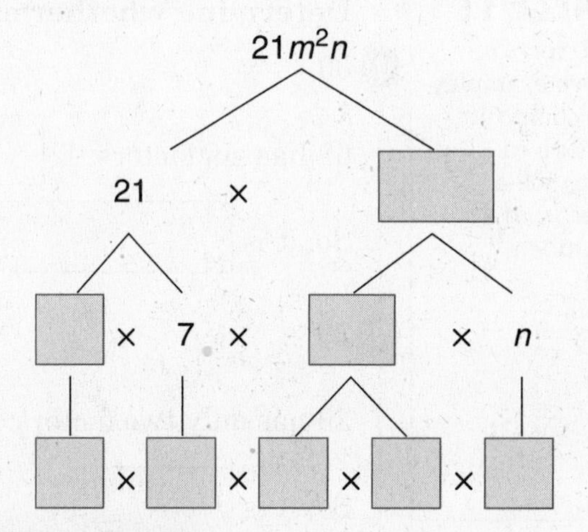

Check Your Progress

a. Find the prime factorization of 72.

b. Factor $15xy^3$.

HOMEWORK ASSIGNMENT

Page(s):

Exercises:

4-2 Greatest Common Factor

MAIN IDEA

- Find the greatest common factor of two or more numbers.

BUILD YOUR VOCABULARY (pages 77–78)

A **Venn diagram** uses ______ to show how elements among sets of numbers or objects are related.

The ______ number that is a common ______ to two or more numbers is called the **greatest common factor (GCF).**

FOLDABLES®

ORGANIZE IT

Under the tab for Lesson 4-2, take notes on finding the greatest common factor of two or more numbers.

EXAMPLE Find the Greatest Common Factor

1 Find the GCF of 28 and 42.

METHOD 1 First, list the factors of 28 and 42.

factors of 28: ______

factors of 42: ______

The common factors are ______.

So, the GCF is ______.

METHOD 2 Use prime factorization.

$28 = 2 \times 2 \times$ ______

$42 = 2 \times 3 \times$ ______

The greatest common factor or GCF is 2×7 or ______.

Check Your Progress Find the GCF of 18 and 45.

EXAMPLE Find the GCF of Three Numbers

WRITE IT

Which method of finding the GCF of two or more numbers do you prefer using to find the GCF of small numbers? for large numbers?

______

______

______

______

______

2 Find the GCF of 21, 42, and 63.

METHOD 1 First, list the factors of 21, 42, and 63.

factors of 21: 1, 3, 7, ____

factors of 42: 1, 2, 3, 6, 7, 14, 21, 42

factors of 63: 1, 3, ____, 9, 21, 63

The common factors of 21, 42, and 63 are ____, ____, and ____.

So, the greatest common factor or GCF is ____.

METHOD 2 Use prime factorization.

$21 = 3 \times 7$

$42 = 2 \times 3 \times 7$ Circle the common factors.

$63 = 3 \times 3 \times 7$

The common prime factors are 3 and 7.

The GCF is ____ × ____, or ____.

Check Your Progress **Find the GCF of each set of numbers.**

24, 48, and 60

EXAMPLE

3 ART **Searra wants to cut a 15-centimeter by 25-centimeter piece of tag board into squares for an art project. She does not want to waste any of the tag board and she wants the largest squares possible. What is the length of the side of the squares she should use?**

The largest length of side possible is the GCF of the dimensions of the tag board.

15 = ☐ × ☐

25 = ☐ × ☐

The ☐ of 15 and 25 is ☐. So, Searra should use ☐ squares with sides measuring ☐ centimeters.

EXAMPLE

4 How many squares can she make if the sides are 5 centimeters?

☐ ÷ 5 = 5 squares can fit along the length.

☐ ÷ 5 = 3 squares can fit along the width.

So, 5 × 3 = ☐ squares can be made from the tag board.

Check Your Progress **CANDY** Alice is making candy baskets using chocolate hearts and lollipops. She is tying each piece of candy with either a red piece of string or a green piece of string. She has 64 inches of red string and 56 inches of green string. She wants to cut the pieces of string equal lengths and use all of the string she has.

a. What is the length of the longest piece of string that can be cut?

b. How many pieces of string can be cut if the pieces are 8 inches long?

HOMEWORK ASSIGNMENT

Page(s):

Exercises:

4-3 Problem-Solving Investigation: Make an Organized List

EXAMPLE Make an Organized List

MAIN IDEA

- Solve problems by making an organized list.

PASSWORD In order to log on to the computer at school, Miranda must use a password. The password is 2 characters. The first character is the letter A or B followed by a single numeric digit. How many passwords does Miranda have to choose from?

UNDERSTAND You know that the password has ______ characters and that the first character is either the letter ______ or B. You know that the second character is a numeric digit. You need to know how many passwords can be created.

PLAN Make an organized list.

SOLVE

A	B	A	B	A	B		B	A	B
0	0		1	2	2	3	3	4	4

A	B	A	B	A		A	B	A	B
5	5		6	7	7	8	8	9	9

There are ______ passwords.

CHECK Draw a tree diagram to check the result.

Check Your Progress **DELI** At a deli, customers can choose from ham or turkey on wheat, rye, or multi-grain bread. How many sandwich possibilities are there?

HOMEWORK ASSIGNMENT

Page(s):

Exercises:

4–4

Simplifying Fractions

MAIN IDEA

- Write fractions in simplest form.

BUILD YOUR VOCABULARY (pages 77–78)

Fractions having the same ______ are called **equivalent fractions.**

A fraction is in **simplest form** when the greatest common factor of the ______ and the denominator is 1.

FOLDABLES

ORGANIZE IT

Under the tab for Lesson 4-4, take notes about simplifying fractions. Be sure to include an example.

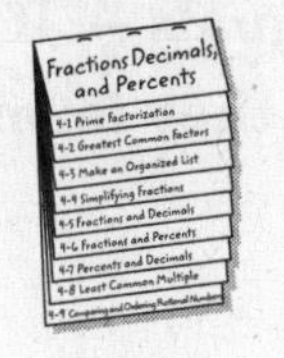

EXAMPLES Write Fractions in Simplest Form

Write each fraction in simplest form.

1 $\frac{12}{45}$

To write a fraction in simplest form, you can divide by common factors or divide by the ______. Let's divide by the GCF.

First, find the GCF of the ______ and ______.

factors of 12: ______

factors of 45: ______

The GCF of 12 and 45 is ______.

Then, divide the numerator and the denominator by ______.

$$\frac{12}{45} = \frac{12 \div \square}{45 \div \square} = \square$$

So, $\frac{12}{45}$ written in simplest form is $\frac{4}{15}$.

2 $\frac{40}{64}$

factors of 40: 1, 2, ☐, 5, 8, 10, 20, ☐

factors of 64: 1, 2, 4, 8, ☐, 32, 64

The GCF of 40 and 64 is ☐.

$$\frac{40}{64} = \frac{40 \div \square}{64 \div \square} = \square$$

So, $\frac{40}{64}$ written in simplest form is ☐.

Check Your Progress **Write each fraction in simplest form.**

a. $\frac{32}{40}$ ☐ **b.** $\frac{28}{49}$ ☐

EXAMPLE

3 **MUSIC Two notes form a *perfect fifth* if the simplified fraction of the frequencies of the notes equals $\frac{3}{4}$. If note D = 294 Hertz and note G = 392 Hertz, do they form a *perfect fifth*?**

$$\frac{\text{frequency of note D}}{\text{frequency of note G}} = \square$$

$$= \frac{\overset{1}{\cancel{2}} \times 3 \times \overset{1}{\cancel{7}} \times \overset{1}{\cancel{7}}}{\underset{1}{\cancel{2}} \times 2 \times 2 \times \underset{1}{\cancel{7}} \times \underset{1}{\cancel{7}}} = \square$$

The fraction of the frequency of the notes D and G is ☐.

So, the two notes do form a *perfect fifth*.

HOMEWORK ASSIGNMENT

Page(s):

Exercises:

Check Your Progress In a bag of 96 marbles, 18 of the marbles are black. Write the fraction of black marbles in simplest form.

4–5 Fractions and Decimals

EXAMPLES Use Mental Math

MAIN IDEA

- Write fractions as terminating or repeating decimals and write decimals as fractions.

Write each fraction or mixed number as a decimal.

1 $\frac{9}{10}$

THINK $\frac{9}{10} = \boxed{}$ (× 10 above and below)

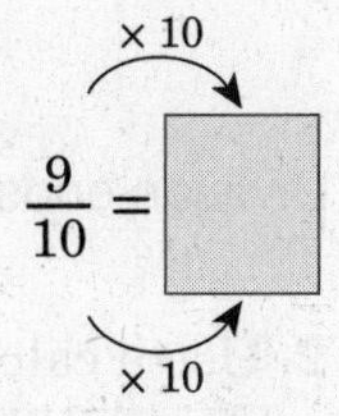

So, $\frac{9}{10} = \boxed{}$.

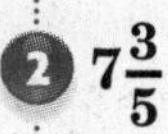

2 $7\frac{3}{5}$

$7\frac{3}{5} = 7 + \boxed{}$ Think of it as a sum.

$= 7 + \boxed{}$ You know that $\frac{3}{5} = 0.6$.

$= 7.6$ Add mentally.

So, $7\frac{3}{5} = \boxed{}$.

Check Your Progress **Write each fraction or mixed number as a decimal.**

a. $\frac{7}{25}$

b. $9\frac{1}{5}$

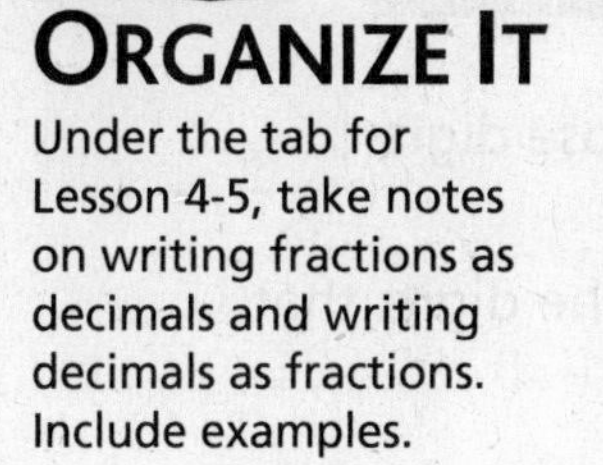

FOLDABLES

ORGANIZE IT

Under the tab for Lesson 4-5, take notes on writing fractions as decimals and writing decimals as fractions. Include examples.

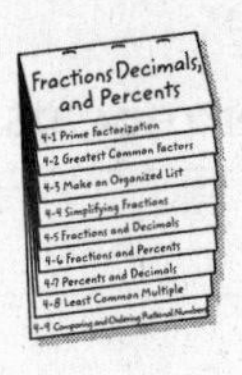

EXAMPLE Use Pencil and Paper or a Calculator

3 Write $\frac{1}{8}$ as a decimal.

METHOD 1 Use paper and pencil.

$$\begin{array}{r} 0.125 \\ 8\overline{)1.000} \\ -\underline{8} \\ 20 \\ -\underline{16} \\ 40 \\ -\underline{40} \\ 0 \end{array}$$

Divide ______ by ______.

Division ends when the remainder is 0.

METHOD 2 Use a calculator.

1 [÷] 8 [ENTER] ______

So, $\frac{1}{8}$ = ______.

Check Your Progress **Write each fraction or mixed number as a decimal.**

a. $\frac{2}{5}$

b. $1\frac{7}{20}$

WRITE IT

Write the following decimal equivalents: $\frac{1}{2}, \frac{1}{3}, \frac{2}{3}, \frac{1}{4}, \frac{3}{4}, \frac{1}{5}, \frac{1}{10}, \frac{1}{8}$.

______

______

______

______

______

BUILD YOUR VOCABULARY (pages 77–78)

A **terminating decimal** is a decimal whose digits ______.

Repeating decimals have a pattern in the digits that repeats ______.

Bar notation is used to indicate that a number repeats forever by writing a ______ over the ______ that repeat.

EXAMPLES Write Fractions as Repeating Decimals

4 Write $\frac{1}{11}$ as a decimal.

METHOD 1 Use paper and pencil.

$$11\overline{)1.0000}$$ with quotient 0.0909...

0
100

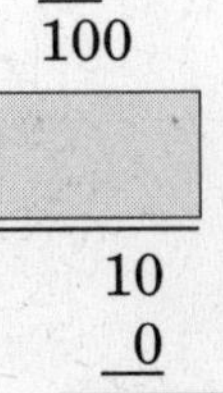

10
0

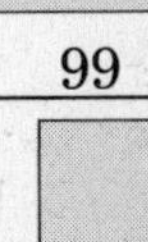

99

METHOD 2 Use a calculator.

1 [÷] 11 [ENTER] 0.0909...

So, $\frac{1}{11}$ = ______.

Check Your Progress Write $2\frac{5}{11}$ as a decimal.

EXAMPLE Use a Power of 10

5 CEREAL Jorge read that 0.72 of his favorite cereal was whole-grain wheat. Find what fraction of his cereal, in simplest form, is whole-grain wheat.

$0.72 = \frac{72}{100}$ The final digit, ______, is in the hundredths place.

$= \frac{18}{25}$ Simplify.

So, ______ of the cereal is whole-grain wheat.

Page(s):

Exercises:

Check Your Progress **EXERCISE** Jeanette ran 0.86 of a mile. What fraction of a mile did she run?

4–6 Fractions and Percents

MAIN IDEA

- Write fractions as percents and percents as fractions.

BUILD YOUR VOCABULARY (pages 77–78)

A **ratio** is a ______ of two numbers by ______.

When a ______ compares a number to ______, it can be written as a **percent.**

KEY CONCEPT

Percent A percent is a ratio that compares a number to 100.

EXAMPLES Write Ratios as Percents

Write each ratio as a percent.

1 Diana scored 63 goals out of 100 attempts.

You can represent 63 out of 100 with a model.

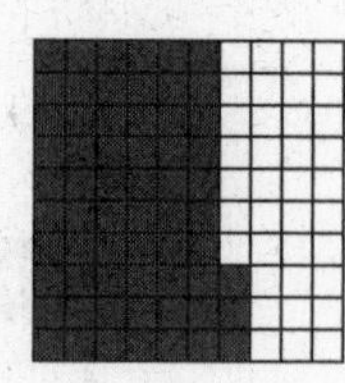

$\frac{63}{100}$ = ______

2 In a survey, 31.9 out of 100 people on average preferred crunchy peanut butter.

$\frac{31.9}{\_\_\_\_}$ = ______

Check Your Progress **Write each ratio as a percent.**

a. Alicia sold 34 of the 100 cookies at the bake sale.

b. On average, 73.4 out of 100 people preferred the chicken instead of the roast beef.

EXAMPLE Write a Fraction as a Percent

3 Write $\frac{16}{25}$ as a percent.

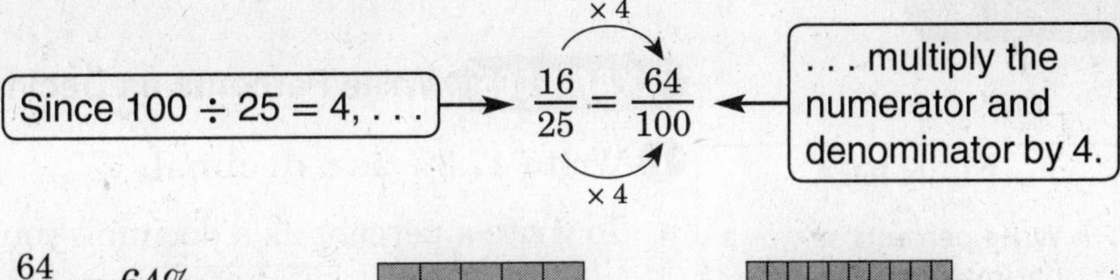

$\frac{64}{100} = 64\%$

So, $\frac{16}{24} = 64\%$.

$\frac{16}{25}$ = $\frac{64}{100}$

Check Your Progress Write $\frac{11}{20}$ as a percent.

FOLDABLES

ORGANIZE IT

Under the tab for Lesson 4-6, take notes on writing fractions as percents and percents as fractions. Include examples.

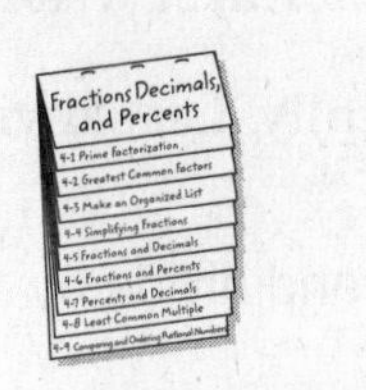

EXAMPLE

4 FISHING William caught and released 20 trout on his fishing trip. Twelve of them were rainbow trout. What percent of the trout he caught were rainbow trout?

William caught ______ rainbow trout out of 20 trout.

$\frac{12}{20}$ = ______ Write an equivalent fraction with a denominator of 100.

$= 60\%$ $\frac{60}{100} = 60\%$

So, ______ of the trout William caught were rainbow trout.

Check Your Progress **READING** Mitchell read 18 out of 25 chapters of a book during his winter vacation. What percent of chapters did he read?

HOMEWORK ASSIGNMENT

Page(s):

Exercises:

4-7 Percents and Decimals

EXAMPLES Write Percents as Decimals

MAIN IDEA

- Write percents as decimals and decimals as percents.

1 **Write 47.8% as a decimal.**

To write a percent as a decimal, you can either first write the percent as a ______ or divide mentally. Let's divide mentally.

$47.8\% = 47.8$	Remove the % symbol and divide by 100.
$= 0.478$	Add leading zero.

So, 47.8% = ______.

2 POPULATION **According to the Administration on Aging, about $28\frac{1}{5}\%$ of the population of the United States is 19 years of age or younger. Write $28\frac{1}{5}\%$ as a decimal.**

$28\frac{1}{5}\% = 28.2\%$	Write $\frac{1}{5}$ as 0.2.
$= 28.2$	Remove the % symbol and divide by 100.
= ______	Add leading zero.

So, $28\frac{1}{5}\% = 0.282$.

KEY CONCEPT

Writing Percents as Decimals To write a percent as a decimal, divide the percent by 100 and remove the percent symbol.

Check Your Progress

a. Write 83.2% as a decimal.

b. AMUSEMENT PARKS A popular amusement park reports that $17\frac{1}{10}\%$ of its visitors will return at least three times during the year. Write $17\frac{1}{10}\%$ as a decimal.

EXAMPLE Write Decimals as Percents

3 Write 0.33 as a percent.

METHOD 1 Write the decimal as a fraction.

$0.33 = \frac{33}{100}$

= ☐ Write the fraction as a percent.

METHOD 2 Multiply mentally.

$0.33 = 33.0$ Multiply by 100.

$= 33\%$ Add the % symbol.

So, 0.33 = ☐.

Check Your Progress Write 0.7 as a percent.

EXAMPLE

4 POPULATION **In 1790, about 0.05 of the population of the United States lived in an urban setting. Write 0.05 as a percent.**

0.05 = ☐ Definition of decimal

= ☐ Definition of 

Check Your Progress In 2000, the population of Illinois had increased by 0.086 from 1990. Write 0.086 as a percent.

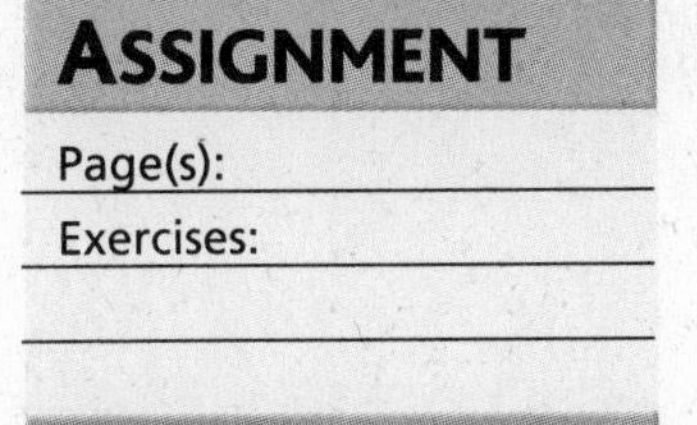

HOMEWORK ASSIGNMENT

Page(s):

Exercises:

4-8 Least Common Multiple

MAIN IDEA

- Find the least common multiple of two or more numbers.

BUILD YOUR VOCABULARY (pages 77–78)

A **multiple** is the ______ of a number and any ______ number.

The **least common multiple (LCM)** of two or more numbers is the ______ of their common multiples, excluding ______.

EXAMPLES Find the LCM

1 Find the LCM of 4 and 6.

METHOD 1 List the nonzero multiples.

multiples of 4:

multiples of 6:

The common multiples are ______, 24, 36, ...

The LCM of 4 and 6 is ______.

METHOD 2 Use prime factorization.

4 = ______ • ______

6 = ______ • ______

The LCM is 2 • 2 • 3 or ______.

FOLDABLES

ORGANIZE IT

Under the tab for Lesson 4-8, take notes about least common multiples. Be sure to include examples.

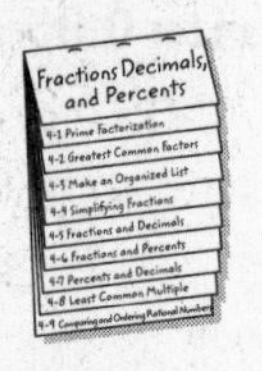

2 **Find the LCM of 4 and 15.**

Use Method 2. Find the prime factorization of each number.

4 = ☐ × 2 or ☐

15 = ☐ × ☐

The prime factors of 4 and 15 are ☐.

The LCM of 4 and 15 is ☐ × 3 × 5, or ☐.

Check Your Progress **Find the LCM of each set of numbers.**

a. 8, 12

b. 6, 14

EXAMPLE

3 **WORK On an assembly line, machine A must be oiled every 18 minutes, machine B every 24 minutes, and machine C every 48 minutes. If all three machines are turned on at the same time, in how many minutes will all three machines need to be oiled at the same time?**

First find the LCM of 18, 24, and 48.

$18 = 2 \times 3 \times 3$ or 2×3^2

$24 = 2 \times 2 \times 2 \times 3$ or $2^3 \times 3$

$48 = 2 \times 2 \times 2 \times 2 \times$ ☐ or $2^4 \times 3$

The LCM of 18, 24, and 48 is $2^4 \times 3^2$ or ☐ × 9, which is 144.

So, all three machines will need to be oiled at the same time in ☐ minutes.

Check Your Progress **LIGHTS** Brenda put up three different strands of decorative blinking lights. The first strand blinks every 6 seconds while the second strand blinks every 8 seconds. The third strand blinks every 4 seconds. If all strands blink at the same time, in how many seconds will they again blink at the same time?

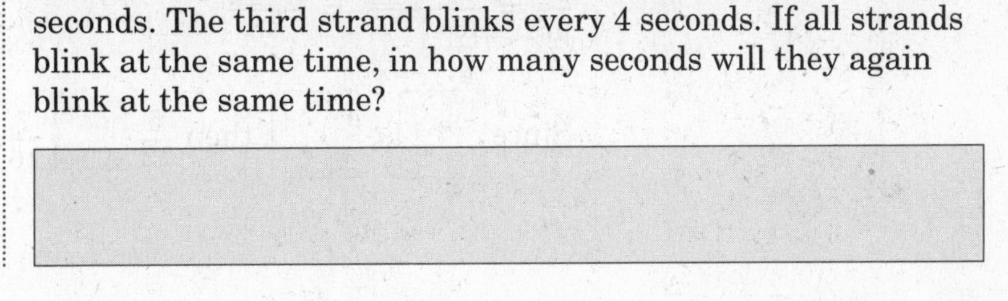

HOMEWORK ASSIGNMENT

Page(s):

Exercises:

4–9 Comparing and Ordering Rational Numbers

MAIN IDEA

- Compare and order fractions, decimals, and percents.

BUILD YOUR VOCABULARY (pages 77–78)

Rational numbers are numbers that can be written as fractions and include fractions, terminating and repeating decimals, and ______.

A **common denominator** is a common multiple of two or more ______.

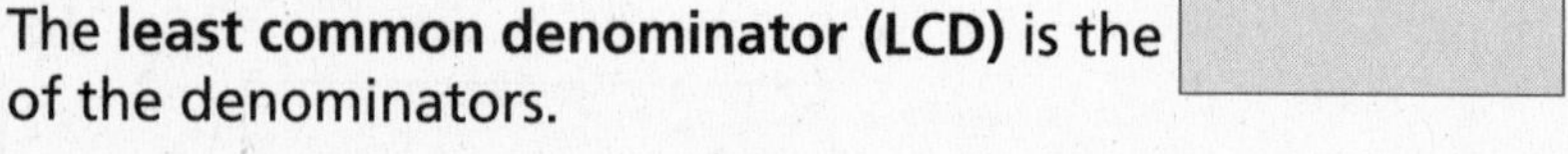

The **least common denominator (LCD)** is the ______ of the denominators.

REVIEW IT

Explain how to write $\frac{48}{60}$ as a decimal. *(Lesson 4-5)*

______

______

______

______

______

EXAMPLES Compare Rational Numbers

Replace each ● with <, >, or = to make a true sentence.

1 $-3\frac{3}{8}$ ● $-3\frac{7}{8}$

Graph each rational number on a number line.

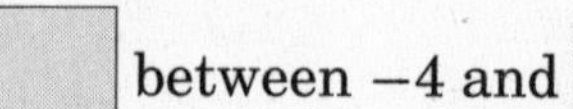

Mark off equal size increments of ______ between −4 and ______.

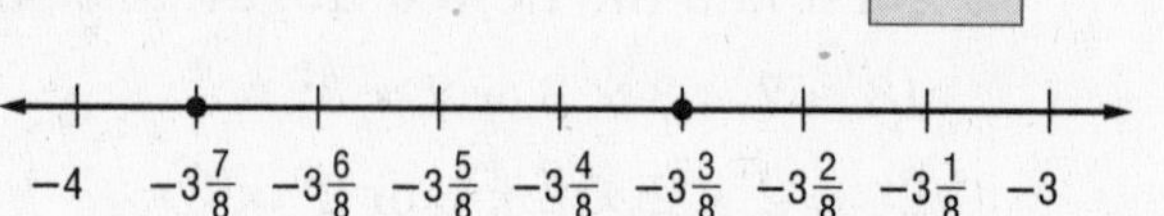

The number line shows that $-3\frac{3}{8}$ ______ $-3\frac{7}{8}$.

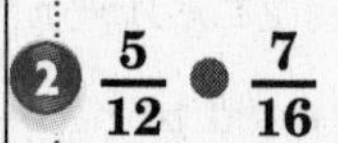

2 $\frac{5}{12}$ ● $\frac{7}{16}$

The LCD of the denominators, 12 and 16, is 48.

$\frac{5}{12} = \frac{5 \cdot \square}{12 \cdot \square} = \frac{\square}{48}$

$\frac{7}{16} = \frac{7 \cdot \square}{16 \cdot \square} = \frac{\square}{48}$

Since ______ < ______, then $\frac{5}{12}$ ______ $\frac{7}{16}$.

Check Your Progress **Replace each ● with $<$, $>$, or $=$ to make a true sentence.**

a. $-2\frac{4}{5}$ ● $-2\frac{3}{5}$

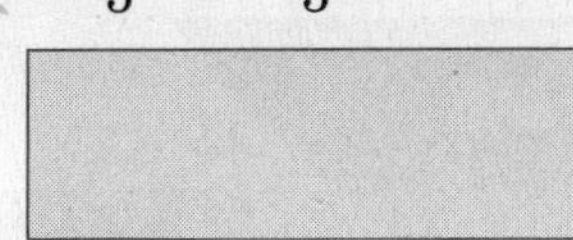

b. $\frac{5}{8}$ ● $\frac{7}{12}$

EXAMPLE

KEY CONCEPT

Rational Numbers Rational numbers are numbers that can be written as fractions.

FOLDABLES Takes notes on rational numbers. Be sure to include examples.

3 DOGS **According to the Pet Food Manufacturer's Association, 11 out of 25 people own large dogs and 13 out of 50 people own medium dogs. Do more people own large or medium dogs?**

Write $\frac{11}{25}$ and $\frac{13}{50}$ as decimals and compare.

$\frac{11}{25} =$ ______ $\frac{13}{50} =$ ______

Since $0.44 > 0.26$, $\frac{11}{25}$ ☐ $\frac{13}{50}$.

So, a greater fraction of people own  dogs than own ______ dogs.

Check Your Progress A survey showed that 21 out of 50 people stated that summer is their favorite season and 13 out of 25 people prefer fall. Do more people prefer summer or fall?

HOMEWORK ASSIGNMENT

Page(s):

Exercises:

CHAPTER 4

BRINGING IT ALL TOGETHER

STUDY GUIDE

FOLDABLES	VOCABULARY PUZZLEMAKER	BUILD YOUR VOCABULARY
Use your **Chapter 4 Foldable** to help you study for your chapter test.	To make a crossword puzzle, word search, or jumble puzzle of the vocabulary words in Chapter 4, go to: glencoe.com	You can use your completed **Vocabulary Builder** (*pages 77–78*) to help you solve the puzzle.

4-1

Prime Factorization

Underline the correct terms to complete each sentence.

1. A factor tree is complete when all of the factors at the bottom of the factor tree are (*prime, composite*) factors.

2. The order of the factors in prime factorization (*does, does not*) matter.

Find the prime factorization of each number.

3. 36

4. 48

5. 250

6. 60

4-2

Greatest Common Factor

Complete each sentence.

7. A ______ shows how elements of sets of numbers are related.

8. A prime factor is a factor that is a ______ number.

9. You can find the ______ of two numbers by ______ the common prime factors.

Find the common prime factors and GCF of each set of numbers.

10. 20, 24 ______

11. 28, 42 ______

4-3 Problem-Solving Investigation: Make an Organized List

12. **CLOTHES** Lucas has a pair of brown pants and a pair of black pants. He has a white dress shirt, a blue dress shirt, and a tan dress shirt. He has a striped tie and a polka-dotted tie. Assuming he can wear any combination, how many combinations of one pair of pants, one dress shirt, and one tie can Lucas wear?

4-4 Simplifying Fractions

Complete the sentence.

13. To find the simplest form of a fraction, ______ the numerator and the denominator by the ______.

Write each fraction in simplest form.

14. $\frac{18}{24}$ ______

15. $\frac{15}{60}$ ______

4-5 Fractions and Decimals

Write each fraction or mixed number as a decimal. Use bar notation if the decimal is a repeating decimal.

16. $3\frac{2}{3}$ ______

17. $5\frac{3}{4}$ ______

18. $\frac{2}{5}$ ______

19. $7\frac{3}{8}$ ______

20. $6\frac{1}{2}$ ______

21. $\frac{7}{10}$ ______

4-6

Fractions and Percents

22. Write the ratio that compares 4 to 25 in three different ways.

23. Write the ratio in exercise 23 as a percent.

24. Write 88% as a fraction in simplest form.

25. Write $\frac{9}{20}$ as a percent.

4-7

Percents and Decimals

Write each percent as a decimal.

26. 69%

27. 3%

28. $32\frac{1}{4}\%$

Write each decimal as a percent.

29. 0.47

30. 0.5775

31. 0.09

4-8

Least Common Multiple

Find the LCM of each set of numbers.

32. 15, 36

33. 21, 70

34. 16, 20

35. 6, 9, 24

36. 12, 18, 30

37. 14, 28, 35

4-9

Comparing and Ordering Rational Numbers

Replace each ● with <, >, or = to make each sentence true.

38. $\frac{14}{35}$ ● $\frac{12}{20}$

39. $\frac{21}{49}$ ● $\frac{18}{63}$

ARE YOU READY FOR THE CHAPTER TEST?

Math Online

Visit **glencoe.com** to access your textbook, more examples, self-check quizzes, and practice tests to help you study the concepts in Chapter 4.

Check the one that applies. Suggestions to help you study are given with each item.

☐ **I completed the review of all or most lessons without using my notes or asking for help.**

- You are probably ready for the Chapter Test.
- You may want to take the Chapter 4 Practice Test on page 225 of your textbook as a final check.

☐ **I used my Foldables or Study Notebook to complete the review of all or most lessons.**

- You should complete the Chapter 4 Study Guide and Review on pages 221–224 of your textbook.
- If you are unsure of any concepts or skills, refer back to the specific lesson(s).
- You may want to take the Chapter 4 Practice Test on page 225 of your textbook.

☐ **I asked for help from someone else to complete the review of all or most lessons.**

- You should review the examples and concepts in your Study Notebook and Chapter 4 Foldables.
- Then complete the Chapter 1 Study Guide and Review on pages 221–224 of your textbook.
- If you are unsure of any concepts or skills, refer back to the specific lesson(s).
- You may also want to take the Chapter 4 Practice Test on page 225 of your textbook.

Student Signature

Parent/Guardian Signature

Teacher Signature

Applying Fractions

Use the instructions below to make a Foldable to help you organize your notes as you study the chapter. You will see Foldable reminders in the margin of this Interactive Study Notebook to help you in taking notes.

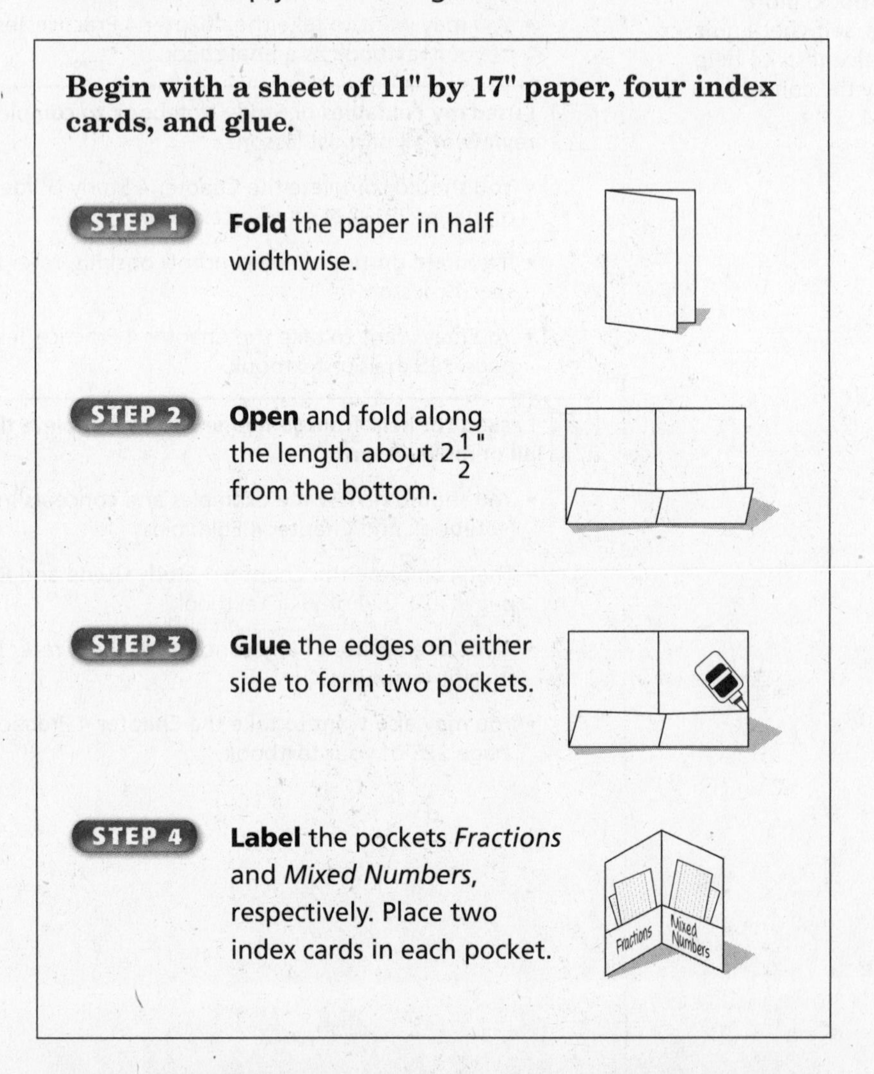

Begin with a sheet of 11" by 17" paper, four index cards, and glue.

STEP 1 **Fold** the paper in half widthwise.

STEP 2 **Open** and fold along the length about $2\frac{1}{2}$" from the bottom.

STEP 3 **Glue** the edges on either side to form two pockets.

STEP 4 **Label** the pockets *Fractions* and *Mixed Numbers*, respectively. Place two index cards in each pocket.

NOTE-TAKING TIP: When you take notes, place a question mark next to any concepts you do not understand. Be sure to ask your teacher to clarify these concepts before a test.

BUILD YOUR VOCABULARY

This is an alphabetical list of new vocabulary terms you will learn in Chapter 5. As you complete the study notes for the chapter, you will see Build Your Vocabulary reminders to complete each term's definition or description on these pages. Remember to add the textbook page number in the second column for reference when you study.

Vocabulary Term	Found on Page	Definition	Description or Example
compatible numbers			
like fractions			
multiplicative inverse [MUHL-tuh-PLIH-kuh-tihv]			
reciprocal [rih-SIH-pruh-kuhl			
unlike fractions			

Chapter 5

5-1 Estimating with Fractions

EXAMPLES Estimate with Mixed Numbers

MAIN IDEA

- Estimate sums, differences, products, and quotients of fractions and mixed numbers.

Estimate.

1 $5\frac{1}{4} + 3\frac{5}{8}$

$5\frac{1}{4} + 3\frac{5}{8} \longrightarrow 5 + \square = \square$

The sum is *about* $\square$.

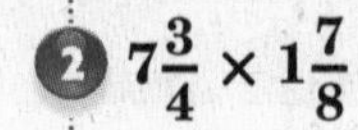

2 $7\frac{3}{4} \times 1\frac{7}{8}$

$7\frac{3}{4} \times 1\frac{7}{8} \longrightarrow \square \times \square = \square$

The sum is *about* $\square$.

Check Your Progress **Estimate.**

a. $2\frac{7}{9} + 5\frac{1}{4}$

b. $4\frac{2}{3} \times 3\frac{1}{8}$

EXAMPLES Estimate with Fractions

Estimate.

3 $\frac{1}{3} + \frac{4}{7}$

$\frac{1}{3}$ is about $\frac{1}{2}$.

$\frac{4}{7}$ is about $\frac{1}{2}$.

$\frac{1}{3} + \frac{4}{7} \longrightarrow \square + \square = \square$

The sum is *about* $\square$.

FOLDABLES

ORGANIZE IT

Record main ideas, definitions and other notes about estimating with fractions on study cards. Store these cards in the "Fractions" pocket of your Foldable.

REMEMBER IT

Some fractions are easy to round because they are close to 1. Examples of these kinds of fractions are ones where the numerator is one less than the denominator, such as $\frac{4}{5}$ or $\frac{7}{8}$.

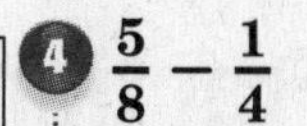

4 $\frac{5}{8} - \frac{1}{4}$

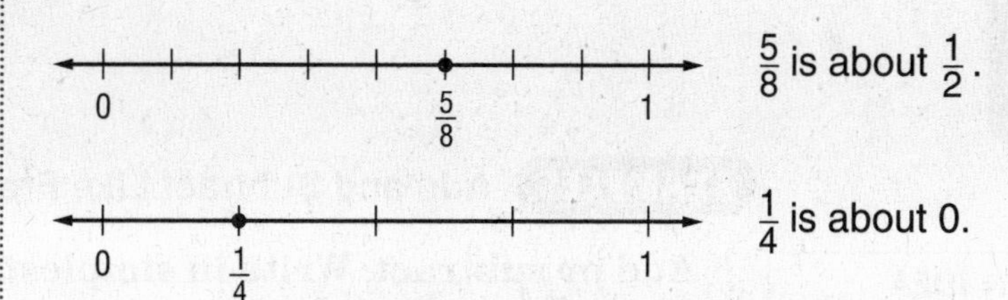

$\frac{5}{8}$ is about $\frac{1}{2}$.

$\frac{1}{4}$ is about 0.

$\frac{5}{8} - \frac{1}{4} \longrightarrow$ ☐ − ☐ = ☐

The difference is *about* ☐.

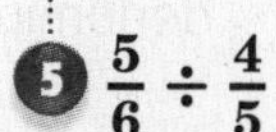

5 $\frac{5}{6} \div \frac{4}{5}$

$\frac{5}{6} \div \frac{3}{4} \approx$ ☐ ÷ ☐ $= 1$

$\frac{5}{6} \approx$ ☐ and $\frac{3}{4} \approx$ ☐.

Check Your Progress **Estimate.**

a. $\frac{8}{9} + \frac{1}{6}$ **b.** $\frac{11}{12} - \frac{2}{9}$ **c.** $\frac{3}{5} \div \frac{7}{8}$

BUILD YOUR VOCABULARY (page 103)

Numbers that are easy to compute ☐ are called **compatible numbers.**

EXAMPLE **Use Compatible Numbers**

6 **Estimate $\frac{3}{4} \times 21$ using compatible numbers.**

$\frac{3}{4} \times 21 \approx \frac{3}{4} \times 20$ or ☐

Round 21 to 20, since 20 is divisible by 4.

Check Your Progress Estimate $\frac{2}{3} \times 17$ using compatible numbers.

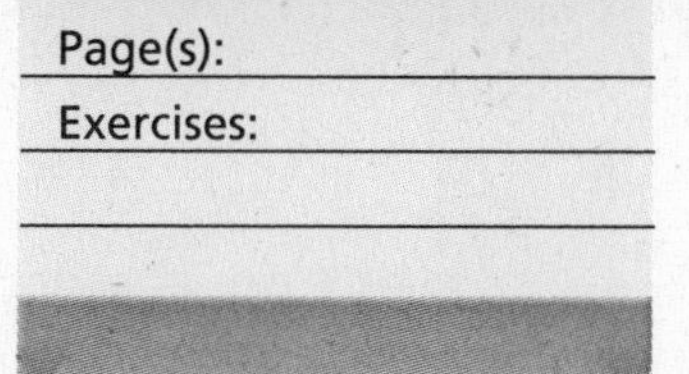

HOMEWORK ASSIGNMENT

Page(s):

Exercises:

5-2 Adding and Subtracting Fractions

MAIN IDEA

- Add and subtract fractions.

KEY CONCEPT

Adding and Subtracting Like Fractions To add or subtract like fractions, add or subtract the numerators and write the result over the denominator. Simplify if necessary.

EXAMPLES Add and Subtract Like Fractions

Add or subtract. Write in simplest form.

1 $\frac{7}{12} + \frac{4}{12}$

$\frac{7}{12} + \frac{4}{12} = \frac{\square}{12}$ Add the ________.

$= \square$ Write the sum over the denominator.

2 $\frac{5}{6} - \frac{1}{6}$

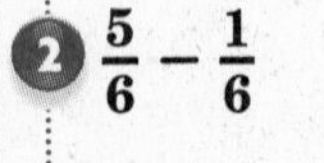

$\frac{5}{6} - \frac{1}{6} = \frac{\square}{6}$ ________ the numerators.

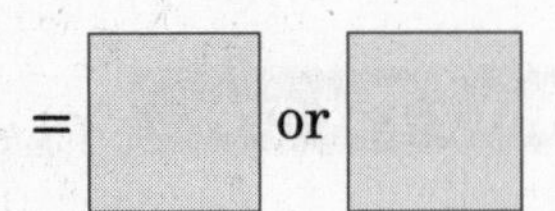

$= \square$ or $\square$ Write the difference sover the

________. Simplify.

EXAMPLES Add and Subtract Unlike Fractions

Add or subtract. Write in simplest form.

3 $\frac{1}{3} + \frac{1}{9}$

To add or subtract unlike fractions, you can use a ________ or

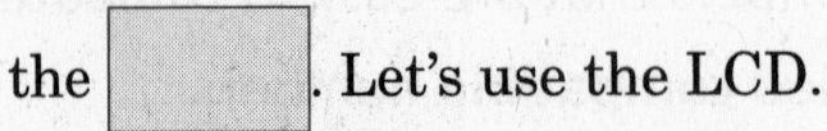

the ________. Let's use the LCD.

The least common denominator of 3 and 9 is ________.

$\frac{1}{3} = \frac{1 \times 3}{\square} = \frac{3}{9}$ Rename $\frac{1}{3}$ using the ________.

$\frac{1}{3} \longrightarrow \square$

$+\frac{1}{9} \longrightarrow +\frac{1}{9}$

$\square$

So, $\frac{1}{3} + \frac{1}{9} = \square$.

WRITE IT

Explain what happens to denominators when adding like fractions.

4 $\frac{3}{4} - \frac{1}{6}$

The LCD of 4 and 6 is ☐.

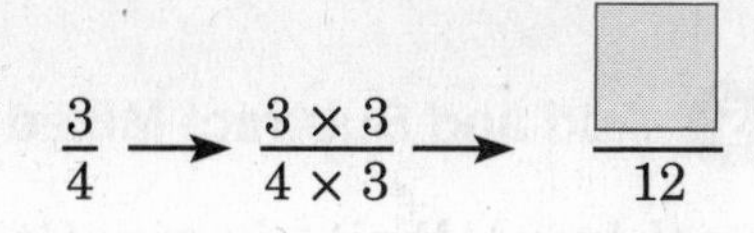

$\frac{3}{4} \rightarrow \frac{3 \times 3}{4 \times 3} \rightarrow \frac{\square}{12}$ Rename each fraction using the LCD.

$-\frac{1}{6} \rightarrow \frac{1 \times 2}{6 \times 2} \rightarrow -\frac{\square}{12}$

$\square$

So, $\frac{3}{4} - \frac{1}{6} = \square$.

Check Your Progress **Add or subtract. Write in simplest form.**

a. $\frac{7}{15} + \frac{4}{15}$ **b.** $\frac{3}{8} + \frac{1}{4}$ **c.** $\frac{7}{9} - \frac{1}{6}$

EXAMPLE

5 ART **A picture mounted on art board is $\frac{1}{8}$ inch thick. The frame for the picture is $\frac{1}{2}$ inch thick. How much thicker than the picture is the frame?**

The phrase *how much thicker* suggests ☐, so find $\frac{1}{2} - \frac{1}{8}$.

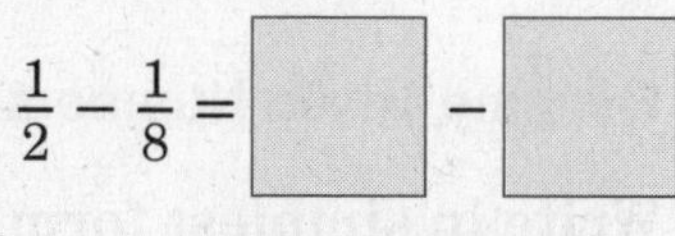

$\frac{1}{2} - \frac{1}{8} = \square - \square$ Rename the fractions using the LCD.

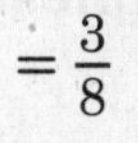

$= \frac{3}{8}$ Subtract the numerators.

The frame is ☐ inch thicker than the picture.

HOMEWORK ASSIGNMENT

Page(s):

Exercises:

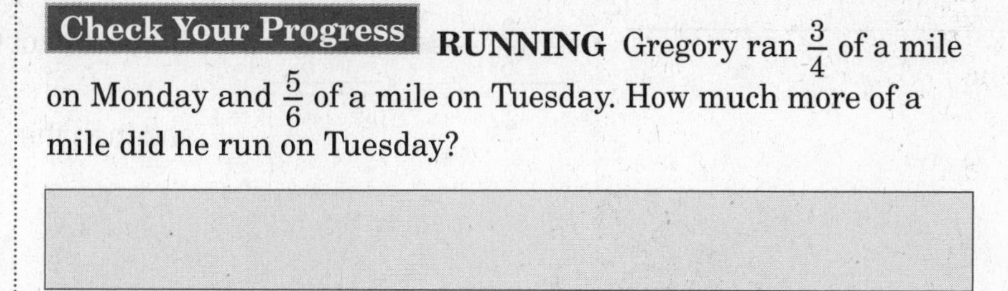

Check Your Progress **RUNNING** Gregory ran $\frac{3}{4}$ of a mile on Monday and $\frac{5}{6}$ of a mile on Tuesday. How much more of a mile did he run on Tuesday?

5-3 Adding and Subtracting Mixed Numbers

EXAMPLES Add and Subtract Mixed Numbers

MAIN IDEA

- Add and subtract mixed numbers.

Add or subtract. Write in simplest form.

1 $3\frac{1}{12} + 14\frac{7}{12}$

Estimate $3 + 15 =$ ____

$$\begin{array}{r} 3\frac{1}{12} \\ +14\frac{7}{12} \\ \hline \end{array}$$

Add the whole numbers and fractions separately.

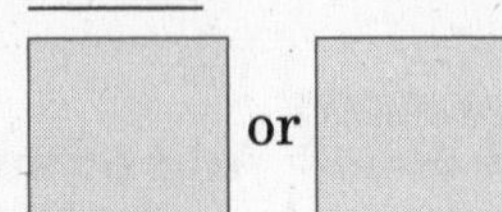

____ or ____

Simplify. Compare the sum to the estimate.

2 $9\frac{7}{10} - 4\frac{3}{5}$

Estimate $10 - 5 =$ ____

$9\frac{7}{10} \longrightarrow 9\frac{7}{10}$

$-4\frac{3}{5} \longrightarrow$ ____

Rename the fraction using the ____.

Simplify. Compare the sum to the estimate.

FOLDABLES

ORGANIZE IT

Record main ideas, definitions, and other notes about adding and subtracting mixed numbers on study cards. Store the cards in the "Mixed Numbers" pocket of your Foldable.

EXAMPLES Rename Mixed Numbers to Subtract

Subtract. Write in simplest form.

3 $8\frac{1}{5} - 3\frac{3}{5}$

$8\frac{1}{5} \longrightarrow 7\frac{6}{5}$

Rename $8\frac{1}{5}$ as ____.

$-3\frac{3}{5} \longrightarrow -3\frac{3}{5}$

First subtract the ____

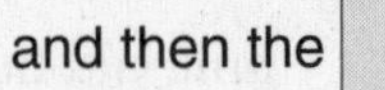

and then the ____.

4 $11 - 8\frac{2}{3}$

$11 \longrightarrow$ ____

$-8\frac{2}{3} \longrightarrow -8\frac{2}{3}$ Subtract.

____

Remember It

When you are adding mixed numbers, you can add the whole numbers first and then add the fractions. Make sure if the fractions add to more than one, that you change the sum of the whole numbers.

Check Your Progress **Add or subtract. Write in simplest form.**

a. $5\frac{5}{14} + 4\frac{3}{14}$ **b.** $6\frac{2}{9} - 3\frac{5}{9}$ **c.** $9\frac{3}{8} - 5\frac{3}{4}$

EXAMPLE

5 **COOKING** **A quiche recipe calls for $2\frac{3}{4}$ cups of grated cheese. A recipe for quesadillas requires $1\frac{1}{3}$ cups of grated cheese. What is the total amount of grated cheese needed for both recipes?**

$2\frac{3}{4} + 1\frac{1}{3} = 2\frac{9}{12} + 1\frac{4}{12}$ Rename the fractions.

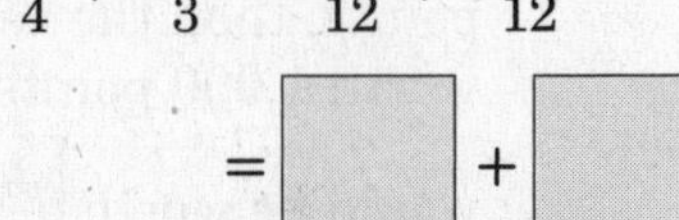

$=$ ____ $+$ ____ Add whole numbers and add fractions.

$= 3 + 1\frac{1}{12}$ or ____ Rename $\frac{13}{12}$ as $1\frac{1}{12}$ and simplify.

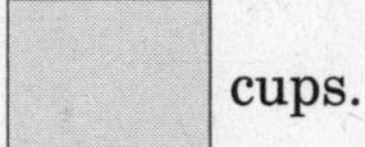

The total amount of grated cheese needed is ____ cups.

Check Your Progress **TIME** Jordan spent $3\frac{1}{6}$ hours at the mall and $2\frac{1}{4}$ hours at the movies. How many more hours did he spend at the mall than at the movies?

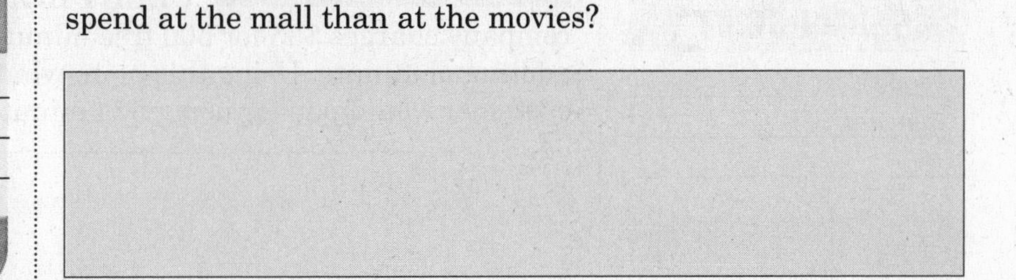

Homework Assignment

Page(s):

Exercises:

5-4 Problem-Solving Investigation: Eliminate Possibilities

EXAMPLE Eliminate Possibilities

MAIN IDEA

- Solve problems by eliminating possibilities.

GAMES On a television game show, the winning contestant must answer three questions correctly to win the grand prize. Each question is worth twice as many points as the question before it. The third question is worth 1,000 points. How much is the first question worth—250, 500, or 2,000 points?

UNDERSTAND You know that there are three questions and each question is worth ______ as many points as the question before it. You know that the third question is worth 1,000 points.

PLAN Eliminate answers that are not ______.

SOLVE The first question cannot be worth 2,000 points since each question after it would have to worth more than 2,000 points, and the third question is only ______ points. So, eliminate that choice. If the first question is worth 500 points, then the second question would be worth 1,000 points and the third question would be worth ______ points. So, eliminate that choice. The reasonable answer is 250 points.

CHECK If the first question is worth 250 points, then the second question would be worth ______ points, and the third question would be worth 1,000 points. So, the answer is correct.

HOMEWORK ASSIGNMENT

Page(s):

Exercises:

Check Your Progress **CELL PHONES** A cell phone company charges $35 for 500 free minutes and $0.50 for each additional minute. Using this plan, what is a reasonable price a customer would pay for using 524 minutes—$32, $40, or $47?

5–5 Multiplying Fractions and Mixed Numbers

MAIN IDEA

- Multiply fractions and mixed numbers.

EXAMPLES Multiply Fractions

Multiply. Write in simplest form.

1 $\frac{1}{8} \times \frac{1}{9}$

$\frac{1}{8} \times \frac{1}{9} =$ ______ ← Multiply the numerators. ← Multiply the denominators.

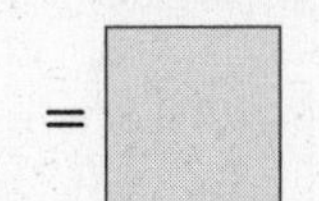

$=$ ______ Simplify.

2 $6 \times \frac{1}{3}$

$6 \times \frac{1}{3} =$ ______ $\times \frac{1}{3}$ Write 6 as ______.

$= \frac{6 \times 1}{1 \times 3}$ Multiply the numerators and the denominators.

$=$ ______ or ______ Simplify.

KEY CONCEPT

Multiplying Fractions
To multiply fractions, multiply the numerators and multiply the denominators.

FOLDABLES Take notes on multiplying fractions and mixed numbers. Place your study cards in your Foldable.

Check Your Progress **Multiply. Write in simplest form.**

a. $\frac{1}{5} \times \frac{1}{7}$

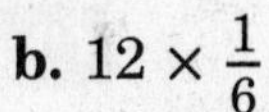

b. $12 \times \frac{1}{6}$

EXAMPLE Simplify Before Multiplying

Multiply. Write in simplest form.

3 $\frac{3}{12} \times \frac{4}{5}$

$\frac{3}{12} \times \frac{4}{5} = \frac{3}{\cancel{12}_{3}} \times \frac{\cancel{4}^{1}}{5}$ Divide 4 and 12 by their GCF, 4.

$= \frac{\_\_\_}{\_\_\_}$ Multiply the numerators and the denominators.

$=$ ______ Simplify.

REMEMBER IT

The Distributive Property can help you do mental math. When you see a problem like $\frac{1}{4} \cdot 4\frac{4}{9}$, you can think, "What is $\frac{1}{4}$ of 4 and what is $\frac{1}{4}$ of $\frac{4}{9}$?" This is equal to $\frac{1}{4}\left(4 + \frac{4}{9}\right)$.

EXAMPLE Multiply Mixed Numbers

4 **Multiply $\frac{1}{3} \times 6\frac{6}{7}$. Write in simplest form.**

METHOD 1 Rename the mixed number.

$\frac{1}{3} \times 6\frac{6}{7} = \frac{\cancel{1}}{\cancel{3}_1} \times \frac{\cancel{48}^{16}}{7}$ Rename $6\frac{6}{7}$ as an ______ fraction, ______.

$= \frac{\boxed{}}{1 \times 7}$ Multiply.

$= \boxed{}$ or $\boxed{}$ Simplify.

METHOD 2 Use mental math.

$\frac{1}{3} \times 6\frac{6}{7} = \frac{1}{3} \times \left(\boxed{} + \boxed{}\right)$ Write $6\frac{6}{7}$ as a sum of its parts.

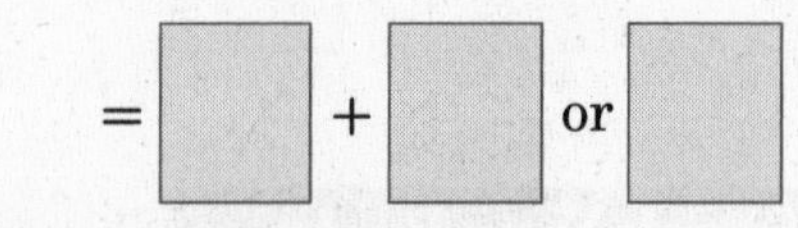

$= \left(\frac{1}{3} \times 6\right) + \left(\frac{1}{3} \times \frac{6}{7}\right)$ ______ Property

$= \boxed{} + \boxed{}$ or $\boxed{}$ Multiply.

Check Your Progress **Multiply. Write in simplest form.**

a. $\frac{4}{9} \times \frac{6}{7}$

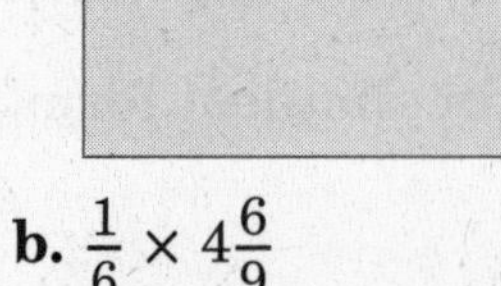

b. $\frac{1}{6} \times 4\frac{6}{9}$

HOMEWORK ASSIGNMENT

Page(s):

Exercises:

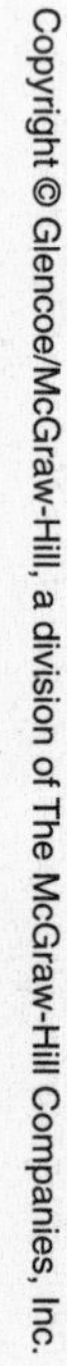

5-6 Algebra: Solving Equations

MAIN IDEA

- Solve equations with rational number solutions.

BUILD YOUR VOCABULARY (page 103)

Two numbers whose ________ is ____ are called **multiplicative inverses.**

Reciprocals is another name given to ________ ________.

KEY CONCEPT

Multiplicative Inverse Property The product of a number and its multiplicative inverse is 1.

EXAMPLES Find Multiplicative Inverses

Find the multiplicative inverse of each number.

1 $\frac{4}{7}$

$\frac{4}{7} \cdot$ ____ $= 1$ — Multiply $\frac{4}{7}$ by ____ to get the product 1.

The multiplicative inverse of $\frac{4}{7}$ is ____, or ____.

2 $6\frac{1}{4}$

$6\frac{1}{4} =$ ____ — Rename the ________ as an improper fraction.

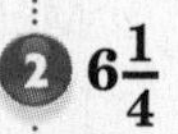

$\frac{25}{4} \cdot$ ____ $= 1$ — Multiply $\frac{25}{4}$ by ____ to get the product 1.

The multiplicative inverse of $6\frac{1}{4}$ is ____.

Check Your Progress **Find the multiplicative inverse of each number.**

a. $\frac{5}{8}$

b. $4\frac{1}{3}$

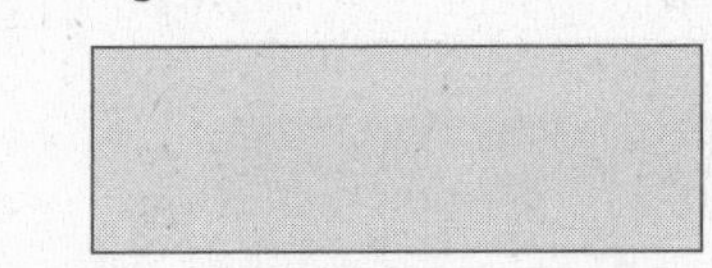

KEY CONCEPT

Multiplication Property of Equality If you multiply each side of an equation by the same nonzero number, the two sides remain equal.

EXAMPLE Solve a Division Equation

3 Solve $11 = \frac{p}{6}$. Check your solution.

$11 = \frac{p}{6}$ Write the equation.

$11 \cdot \square = \frac{p}{6} \cdot \square$ Multiply each side by .

$\square = p$ Simplify.

Check

$11 = \frac{p}{6}$ Write the original equation.

$11 = \frac{\square}{6}$ Replace p with $\square$.

$11 = \square$ Simplify.

The solution is 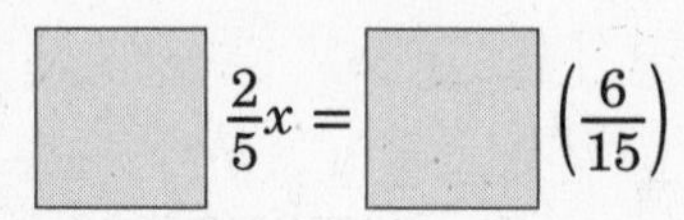.

EXAMPLE Use a Reciprocal to Solve an Equation

4 Solve $\frac{2}{5}x = \frac{6}{15}$.

$\frac{2}{5}x = \frac{6}{15}$ Write the equation.

$\square \frac{2}{5}x = \square \left(\frac{6}{15}\right)$ Multiply each side by the  of $\frac{2}{5}$.

$x = \square$ or $\square$ Simplify.

Check Your Progress **Solve.**

a. $\frac{m}{9} = 4$

b. $\frac{3}{8}x = \frac{3}{4}$

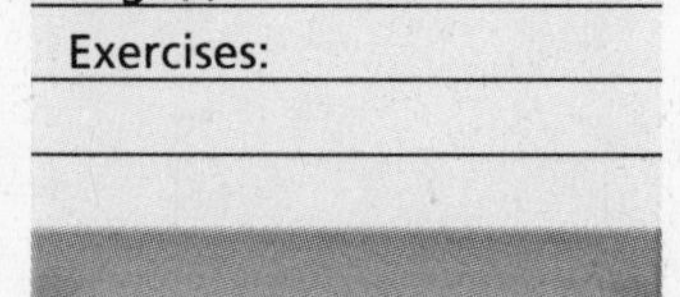

HOMEWORK ASSIGNMENT

Page(s):

Exercises:

5–7 Dividing Fractions and Mixed Numbers

MAIN IDEA

- Divide fractions and mixed numbers.

KEY CONCEPT

Division by a Fraction
To divide by a fraction, multiply by its multiplicative inverse or reciprocal.

EXAMPLE Divide by a Fraction

1 **Find $\frac{2}{3} \div \frac{4}{9}$. Write in simplest form.**

$\frac{2}{3} \div \frac{4}{9} = \frac{2}{3} \cdot$ ____ Multiply by the reciprocal $\frac{4}{9}$.

$= \frac{\overset{1}{\cancel{2}}}{\underset{1}{\cancel{3}}} \cdot \frac{\overset{3}{\cancel{9}}}{\underset{2}{\cancel{4}}}$ Divide out common factors.

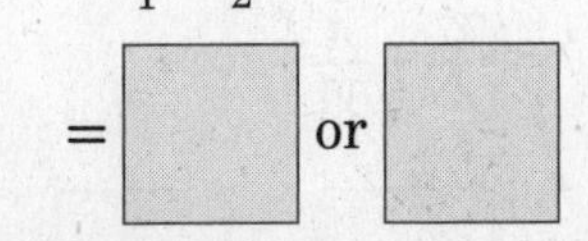

= ____ or ____ Multiply and simplify.

EXAMPLE Divide by Mixed Numbers

2 **Find $\frac{5}{6} \div 2\frac{1}{2}$. Write in simplest form.**

Estimate $1 \div \frac{5}{2} = 1 \times$ ____ or $\frac{2}{5}$

$\frac{5}{6} \div 2\frac{1}{2} = \frac{5}{6} \div$ ____ Rename $2\frac{1}{2}$ as an ____ fraction.

$= \frac{5}{6} \cdot$ ____ Multiply by the reciprocal of $\frac{5}{2}$.

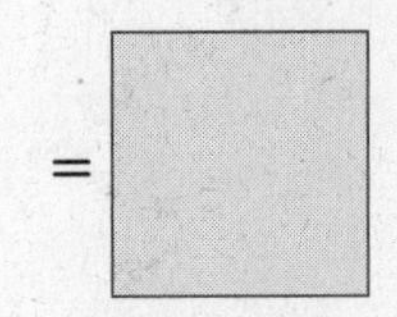

= ____ Divide out common factors.

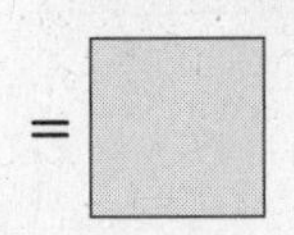

= ____ Multiply. The quotient is close to the estimate.

WRITE IT

Will the quotient $7\frac{1}{6} \div 3\frac{2}{3}$ be a fraction less than 1 or greater than 1? Explain.

Check Your Progress **Divide. Write in simplest form.**

a. $\frac{6}{7} \div \frac{2}{5}$

b. $\frac{3}{8} \div 2\frac{1}{2}$

EXAMPLE

3 **FACTORY A bottling machine needs to be restocked with new lids every $2\frac{3}{4}$ hours. If the machine runs $19\frac{1}{4}$ hours, how many times will it have to be restocked with lids?**

$19\frac{1}{4} \div 2\frac{3}{4} = \square \div \square$ Rename the mixed numbers as improper fractions.

$= \frac{77}{4} \cdot \frac{4}{11}$ Multiply by the ______ of $\frac{11}{4}$, which is $\frac{4}{11}$.

$= \frac{\overset{7}{\cancel{77}}}{\underset{1}{\cancel{4}}} \cdot \frac{\overset{1}{\cancel{4}}}{\underset{1}{\cancel{11}}}$ Divide out common factors.

$= \square$ or $\square$ Multiply.

So, the machine will need to restocked ______ times.

Check Your Progress **FURNITURE** A rectangular table is $5\frac{5}{6}$ feet long. If the area of the table is $20\frac{5}{12}$ square feet, how wide is the table?

HOMEWORK ASSIGNMENT

Page(s):

Exercises:

CHAPTER 5

BRINGING IT ALL TOGETHER

STUDY GUIDE

FOLDABLES	VOCABULARY PUZZLEMAKER	BUILD YOUR VOCABULARY
Use your **Chapter 5 Foldable** to help you study for your chapter test.	To make a crossword puzzle, word search, or jumble puzzle of the vocabulary words in Chapter 5, go to: glencoe.com	You can use your completed **Vocabulary Builder** (*page 103*) to help you solve the puzzle.

5-1 Estimating with Fractions

Estimate.

1. $8\frac{2}{3} + 7\frac{1}{4}$

2. $11\frac{7}{8} \div 3\frac{5}{6}$

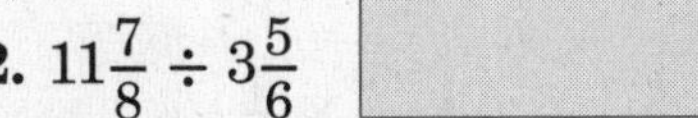

5-2 Adding and Subtracting Fractions

Add or subtract. Write in simplest form.

3. $\frac{7}{8} + \frac{3}{8}$

4. $\frac{5}{6} - \frac{1}{3}$

5. $\frac{1}{5} + \frac{3}{4}$

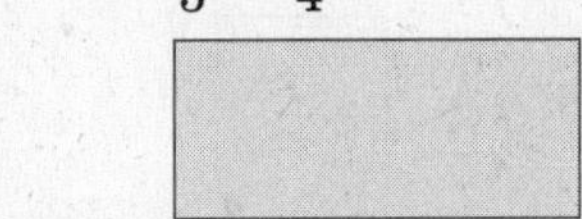

5-3 Adding and Subtracting Mixed Numbers

Add or subtract. Write in simplest form.

6. $3\frac{7}{8} + 6\frac{1}{4}$

7. $7\frac{1}{6} + 2\frac{5}{12}$

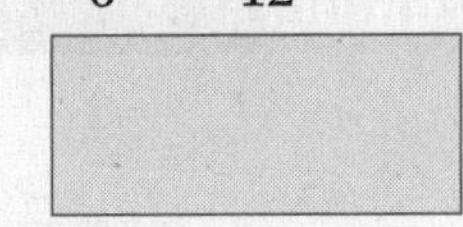

8. $8\frac{3}{7} - 4\frac{5}{7}$

9. $9\frac{2}{9} - 1\frac{2}{3}$

5-4 Problem-Solving Investigation: Eliminate Possibilities

10. READING Joel read $\frac{5}{8}$ of a novel. If the novel has 600 pages, is 250, 300, or 375 a reasonable number of pages that Joel has read?

5-5 Multiplying Fractions and Mixed Numbers

Multiply. Write in simplest form.

11. $\frac{2}{7} \times 4\frac{1}{5}$

12. $\frac{1}{6} \times \frac{3}{4}$

13. $5\frac{1}{6} \times \frac{2}{5}$

14. $\frac{5}{8} \times \frac{4}{5}$

5-6 Algebra: Solving Equations

Find the multiplicative inverse of each number.

15. $\frac{3}{5}$

16. $1\frac{1}{2}$

17. 3

Solve each equation.

18. $\frac{1}{3}a = \frac{5}{6}$

19. $-4 = \frac{k}{3}$

5-7 Dividing Fractions and Mixed Numbers

Divide. Write in simplest form.

20. $\frac{1}{4} \div \frac{2}{3}$

21. $\frac{7}{8} \div \frac{2}{3}$

22. $6 \div 1\frac{1}{3}$

23. $5\frac{3}{4} \div 2\frac{1}{2}$

ARE YOU READY FOR THE CHAPTER TEST?

Math Online

Visit **glencoe.com** to access your textbook, more examples, self-check quizzes, and practice tests to help you study the concepts in Chapter 5.

Check the one that applies. Suggestions to help you study are given with each item.

☐ **I completed the review of all or most lessons without using my notes or asking for help.**

- You are probably ready for the Chapter Test.
- You may want to take the Chapter 5 Practice Test on page 275 of your textbook as a final check.

☐ **I used my Foldable or Study Notebook to complete the review of all or most lessons.**

- You should complete the Chapter 5 Study Guide and Review on pages 271–274 of your textbook.
- If you are unsure of any concepts or skills, refer back to the specific lesson(s).
- You may also want to take the Chapter 5 Practice Test on page 275 of your textbook.

☐ **I asked for help from someone else to complete the review of all or most lessons.**

- You should review the examples and concepts in your Study Notebook and Chapter 5 Foldable.
- Then complete the Chapter 5 Study Guide and Review on pages 271–274 of your textbook.
- If you are unsure of any concepts or skills, refer back to the specific lesson(s).
- You may also want to take the Chapter 5 Practice Test on page 275 of your textbook.

Student Signature

Parent/Guardian Signature

Teacher Signature

Ratios and Proportions

Use the instructions below to make a Foldable to help you organize your notes as you study the chapter. You will see Foldable reminders in the margin of this Interactive Study Notebook to help you in taking notes.

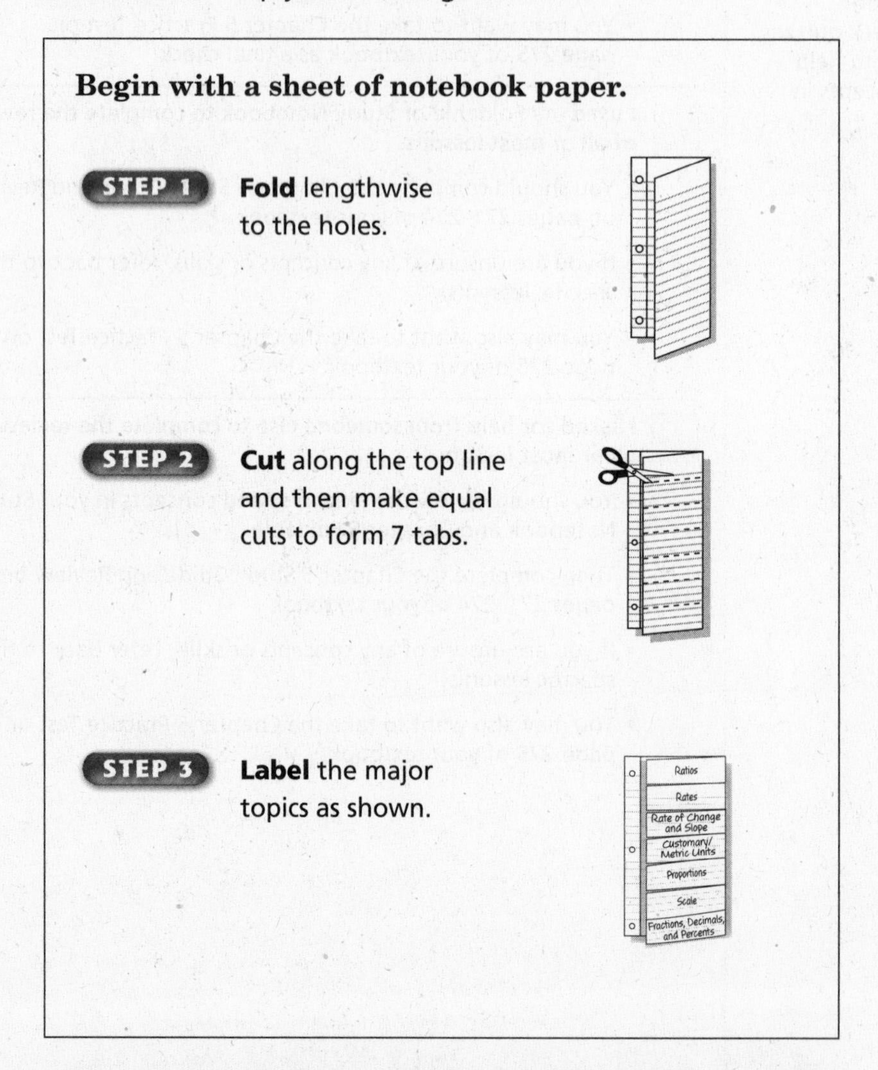

Begin with a sheet of notebook paper.

STEP 1 **Fold** lengthwise to the holes.

STEP 2 **Cut** along the top line and then make equal cuts to form 7 tabs.

STEP 3 **Label** the major topics as shown.

NOTE-TAKING TIP: When you take notes, it may be helpful to include an example for each term or concept learned.

BUILD YOUR VOCABULARY

This is an alphabetical list of new vocabulary terms you will learn in Chapter 6. As you complete the study notes for the chapter, you will see Build Your Vocabulary reminders to complete each term's definition or description on these pages. Remember to add the textbook page number in the second column for reference when you study.

Vocabulary Term	Found on Page	Definition	Description or Example
cross products			
equivalent ratios			
gram			
kilogram			
liter			
meter			
metric system			
proportion			
proportional			
rate			

Vocabulary Term	Found on Page	Definition	Description or Example
ratio			
scale			
scale drawing			
scale factor			
scale model			
slope			
unit rate			
unit ratio			

6–1 Ratios

MAIN IDEA

- Write ratios as fractions in simplest form and determine whether two ratios are equivalent.

BUILD YOUR VOCABULARY (pages 121–122)

A ______ is a comparison of two quantities by division.

Ratios that express the ______ relationship between two quantities are **equivalent ratios**.

EXAMPLE Write Ratios in Simplest Form

1 APPLES **Mr. Gale bought a basket of apples. Using the table, write a ratio comparing the Red Delicious apples to the Granny Smith apples as a fraction in simplest form.**

Mr. Gale's Apples
12 Fuji
9 Granny Smith
30 Red Delicious

$\frac{\text{Red Delicious}}{\text{Granny Smith}} \quad \frac{30}{9} = \frac{\overset{10}{\cancel{30}}}{\underset{3}{\cancel{9}}}$ or ______

The ratio of Red Delicious apples to Granny Smith apples is ______.

EXAMPLE Identify Equivalent Ratios

2 **Determine whether the ratios 12 onions to 15 potatoes and 32 onions to 40 potatoes are equivalent.**

$\text{12 onions} : \text{15 potatoes} = \frac{12 \div 3}{15 \div 3}$ or ______

$\text{32 onions} : \text{40 potatoes} = \frac{32 \div 8}{40 \div 8}$ or ______

The ratios simplify to the same fraction. They are ______.

FOLDABLES

ORGANIZE IT

Record a term or concept from Lesson 6–1 under the Ratios tab and write a definition along with an example to the right of the definition.

Ratios
Rates
Rate of Change and Slope
Customary/Metric Units
Proportions
Scale
Fractions, Decimals, and Percents

Check Your Progress

a. FLOWERS A garden has 18 roses and 24 tulips. Write a ratio comparing roses to tulips as a fraction in simplest form.

b. Determine whether the ratios 3 cups vinegar to 8 cups water and 5 cups vinegar to 12 cups water are equivalent.

REMEMBER IT

Ratios such as 120:1,800 can also be written in simplest form as 1:15.

EXAMPLE

3 POOLS It is recommended that no more than one person be allowed into the shallow end of an outdoor public pool for every 15 square feet of surface area. If a local pool's shallow end has a surface area of 1,800 square feet, are the lifeguards correct to allow 120 people into that part of the pool?

Recommended Ratio

1:15 = ______ persons per square feet

Actual Ratio

120:1,800 = $\frac{120}{1,800}$ or ______ persons per square feet

Since the ratios simplify to the same fraction, they are ______. The lifeguards are correct.

Check Your Progress **SCHOOL** A district claims that they have 1 teacher for every 15 students. If they actually have 2,700 students and 135 teachers, is their claim correct?

HOMEWORK ASSIGNMENT

Page(s):

Exercises:

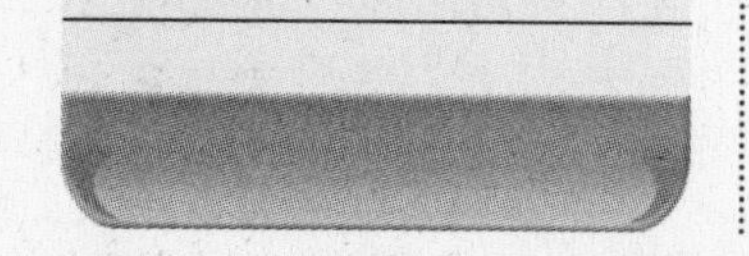

6–2 Rates

BUILD YOUR VOCABULARY (pages 121–122)

A ratio that ________ two quantities with different kinds of units is called a **rate**.

When a rate is simplified so that it has a ________ of 1 unit, it is called a **unit rate**.

MAIN IDEA

- Determine unit rates.

FOLDABLES

ORGANIZE IT

Under the rate tab, take notes on rate and unit rate. Be sure to include examples.

Ratios
Rates
Rate of Change and Slope
Customary/Metric Units
Proportions
Scale
Fractions, Decimals, and Percents

EXAMPLES Find Unit Rates

1 READING Julia read 52 pages in 2 hours. What is the average number of pages she read per hour?

Write the rate as a fraction. Then find an equivalent rate with a denominator of 1.

52 pages in 2 hours $= \frac{52 \text{ pages}}{2 \text{ hours}}$ Write the rate as a fraction.

$= \frac{52 \text{ pages} \div \square}{2 \text{ hours} \div \square}$ Divide the numerator and denominator by ____.

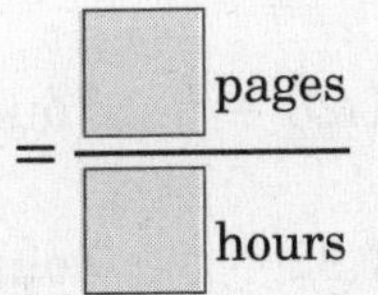

$= \frac{\square \text{ pages}}{\square \text{ hours}}$ Simplify.

2 SODA Find the unit price per can if it costs $3 for 6 cans of soda. Round to the nearest hundredth if necessary.

$3 for 6 cans $= \frac{\$3}{6 \text{ cans}}$ Write the rate as a fraction.

$= \frac{\$3 \div 6}{6 \text{ cans} \div 6}$ Divide the numerator and the denominator by 6.

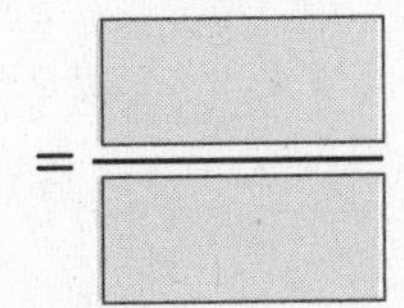

$= \frac{\square}{\square}$ Simplify.

REMEMBER IT

The word *rate* is often understood to mean unit rate.

Check Your Progress **Find each unit rate.**

a. 16 laps in 4 minutes

b. $3 for one dozen cookies

EXAMPLE Compare Using Unit Rates

3 TEST EXAMPLE **The costs of 4 different sizes of orange juice are shown in the table. Which container costs the least per ounce?**

Amount	Total Cost
16 oz	$1.28
32 oz	$1.92
64 oz	$2.56
96 oz	$3.36

A 96-oz container

B 64-oz container

C 32-oz container

D 16-oz container

Read the Item

Find the unit price, or the cost per ounce of each size of orange juice. Divide the price by the number of ounces.

Solve the Item

$1.28 ÷ ______ ounces = ______ per ounce.

$1.92 ÷ ______ ounces = ______ per ounce.

$2.56 ÷ ______ ounces = ______ per ounce.

$3.36 ÷ ______ ounces = ______ per ounce.

The ______ -ounce container of orange juice costs the least per ounce. The answer is ______.

Check Your Progress

MULTIPLE CHOICE The costs of different sizes of bottles of laundry detergent are shown below. Which bottle costs the least per ounce?

Amount	Total Cost
16 oz	$3.12
32 oz	$5.04
64 oz	$7.04
96 oz	$11.52

F 96-oz container

G 64-oz container

H 32-oz container

J 16-oz container

EXAMPLE Use a Unit Rate

4 POTATOES An assistant cook peeled 18 potatoes in 6 minutes. At this rate, how many potatoes can he peel in 50 minutes?

Find the unit rate.

18 potatoes in 6 minutes $= \frac{18 \div 6}{6 \div 6} = \frac{3}{1}$

The unit rate is ______ potatoes per minute.

$\frac{3 \text{ potatoes}}{1 \cancel{\text{min}}} \cdot 50 \cancel{\text{min}} =$ ______ potatoes

He can peel ______ potatoes in 50 minutes.

Check Your Progress Sarah can paint 21 beads in 7 minutes. At this rate, how many beads can she paint in one hour?

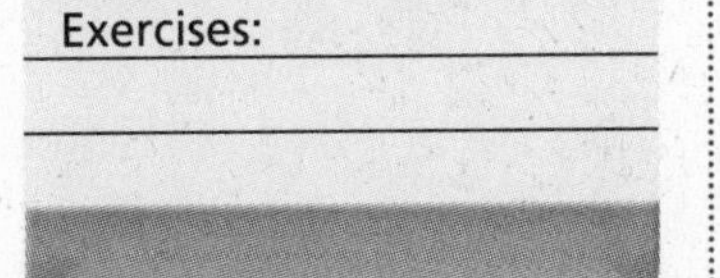

Page(s):

Exercises:

6–3 A Plan for Problem Solving

MAIN IDEA

- Identify rate of change and slope using tables and graphs.

BUILD YOUR VOCABULARY (pages 121–122)

A **rate of change** is a rate that describes how one quantity changes in relation to another and is usually expressed as a ______.

EXAMPLE Find Rate of Change from a Table

1 The table shows the number of miles a car drove on a trip. Use the information to find the approximate rate of change.

+ 65 + 65 + ______

Distance (miles)	65	130	195	260
Time (hours)	1	2	3	4

+ 1 + 1 + 1

$\frac{\text{change in distance}}{\text{change in time}} = \frac{\square}{\square}$ The distance increased ______ miles for every hour.

So, the rate was 65 miles per hour.

Check Your Progress The table shows the number of miles a car drove on a trip. Use the information to find the rate of change.

Distance (miles)	44	88	132	176
Fuel (gallons)	2	4	6	8

WRITE IT

Explain how rate of change is similar to unit rates.

______

______

______

______

BUILD YOUR VOCABULARY (pages 121–122)

The constant rate of change in y with respect to the constant change in ______ is called the **slope** of a line.

EXAMPLE Find Rate of Change from a Graph

FOLDABLES

ORGANIZE IT

Under the rate of change and slope tab, take notes on how to find the slope of a line.

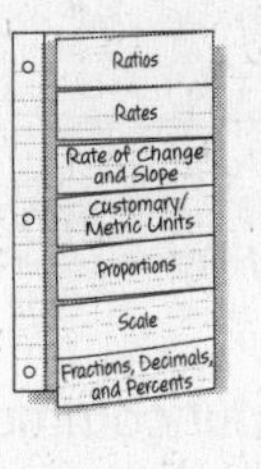

2 GRAPH THE DATA **Find the slope of the line. Explain what the slope represents.**

Graph the points and connect them with a line.

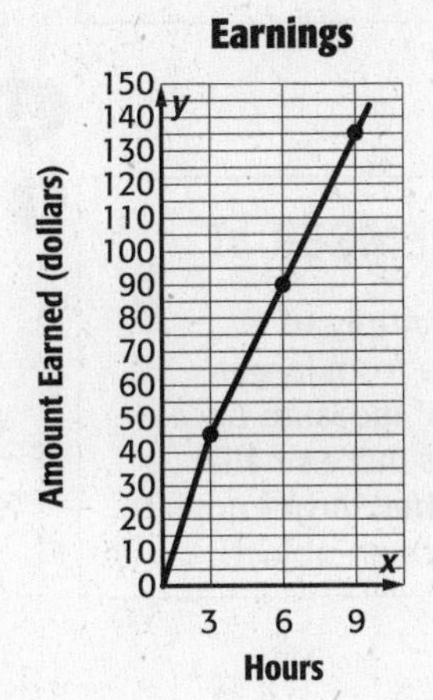

Hours	Amount Earned
3	$45
6	$90
9	$135

Pick two points on the line, such as (3, 45) and (6, 90), to find the slope.

$$\text{slope} = \frac{\text{change in } y}{\text{change in } x}$$

$$= \frac{90 - \boxed{}}{6 - \boxed{}}$$

$$= \frac{45}{3} \text{ or } \boxed{}$$

The slope is ______ and represents the amount earned per hour.

Check Your Progress The table shows the cost of renting a bicycle. Graph the data. Find the slope of the line. Explain what the slope represents.

Hours	Cost
2	$8
4	$16
6	$24

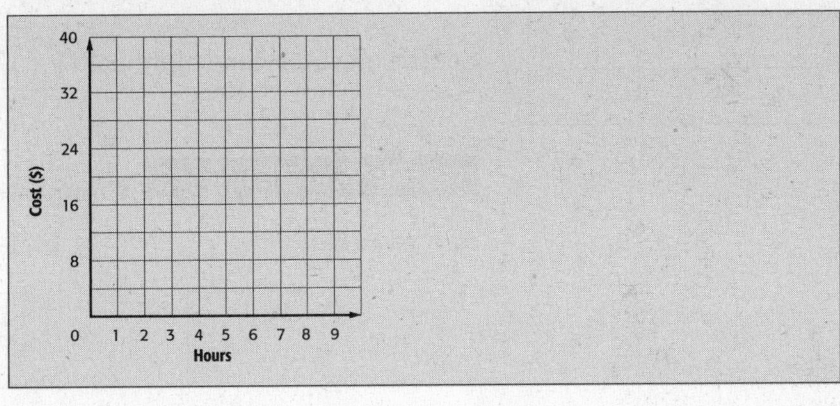

6–4 Measurement: Changing Customary Units

BUILD YOUR VOCABULARY (pages 121–122)

A **unit ratio** is a ratio in which the denominator is ____ unit.

MAIN IDEA

- Change units in the customary system.

REMEMBER IT

You multiply to change from larger units of measure because it takes more smaller units than larger units to measure an object.

REVIEW IT

Explain how estimating can help you solve a problem. *(Lesson 6-1)*

____

____

____

____

____

EXAMPLES Convert Larger Units to Smaller Units

1 Convert 2 miles into feet.

Since 1 mile = 5,280 feet, the unit ratio is .

$2 \text{ mi} = 2 \text{ mi} \cdot \frac{5{,}280 \text{ ft}}{1 \text{ mi}}$ — Multiply by $\frac{5{,}280 \text{ ft}}{1 \text{ mi}}$.

$= 2 \cancel{\text{mi}} \cdot \frac{5{,}280 \text{ ft}}{1 \cancel{\text{mi}}}$ — Divide out common units.

$=$ ____ ft or 10,560 ft — Multiply.

So, 2 miles = ____ feet.

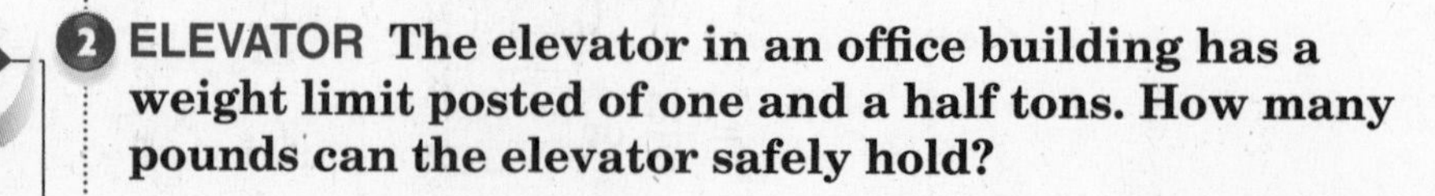

2 ELEVATOR The elevator in an office building has a weight limit posted of one and a half tons. How many pounds can the elevator safely hold?

$1\frac{1}{2} \text{ t} = 1\frac{1}{2} \cancel{\text{t}} \cdot$ 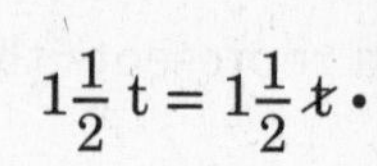 — Multiply by  since there are ____ pounds in 1 ton.

$= 1\frac{1}{2} \cdot 2{,}000 \text{ lb or } 3{,}000 \text{ lb}$ — Multiply.

So, the elevator can safely hold  pounds.

Check Your Progress **Complete.**

a. 8 yd = ■ ft

b. $4\frac{1}{2}$ T = ■ lb

EXAMPLES Convert Smaller Units to Larger Units

3 Convert 11 cups into pints.

Since 1 pint = 2 cups, the unit ratio is $\frac{2\text{ c}}{1\text{ pt}}$, and its reciprocal is ______.

$11\text{ c} = 11\text{ c} \cdot \frac{1\text{ pt}}{2\text{ c}}$ Multiply by ______.

$= 11\not{c} \cdot \frac{1\text{ pt}}{2\not{c}}$ Divide out common units.

$= 11 \cdot$ ______

$= \frac{11}{2}\text{ pt}$ Multiplying 11 by $\frac{1}{2}$ is the same as dividing 11 by 2.

$=$ ______ pt

So, 11 cups = ______ pints.

4 SOCCER Tracy kicked a soccer ball 1,000 inches. How many feet did she kick the ball?

Since 1 foot = 12 inches, multiply by ______. Then divide out common units.

$1{,}000\text{ in.} = 1{,}000\not{\text{in.}} \cdot \frac{1\text{ ft}}{12\not{\text{in.}}}$

$= 1{,}000\text{ in.} \cdot$ ______ ft

$= \frac{1000}{12}\text{ ft or}$ ______ ft

So, Tracy kicked the soccer ball ______.

Check Your Progress **Complete.**

a. 21 qt = ■ gal

b. 78 oz = ■ lb

EXAMPLE

5 **LEMONADE Paul made 6 pints of lemonade and poured it into 10 glasses equally. How many cups of lemonade did each glass contain?**

Begin by converting 6 pints to cups.

$$6 \text{ pt} = 6 \text{ pt} \cdot \frac{\square}{1 \text{ pt}}$$

$$= 6 \cdot 2 \text{ cups or } \square \text{ cups}$$

Find the unit rate which gives the number of cups per glass.

$$\frac{12 \text{ cups}}{10 \text{ glasses}} = \frac{6}{5} \text{ or } \square \text{ cups per glass}$$

Check Your Progress **CANDY** Tom has 3 pounds of candy he plans to divide evenly among himself and his 3 best friends. How many ounces of candy will each of them get?

HOMEWORK ASSIGNMENT

Page(s):

Exercises:

6–5 Measurement: Changing Metric Units

Main Idea

- Change metric units of length, capacity, and mass.

BUILD YOUR VOCABULARY (pages 121–122)

The **metric system** is a ____ system of measures.

The **meter** is the base unit of ____.

The **liter** is the base unit of ____.

The **gram** measures ____.

The base unit of mass in the metric system is the ____.

EXAMPLES Convert Units in the Metric System

1 Complete 7.2 m = ■ mm.

To convert from meters to millimeters, ____ by ____.

7.2 × ____ = ____

So, 7.2 m = ____ mm.

2 Complete 40 cm = ■ m.

To convert from centimeters to meters, ____ by ____.

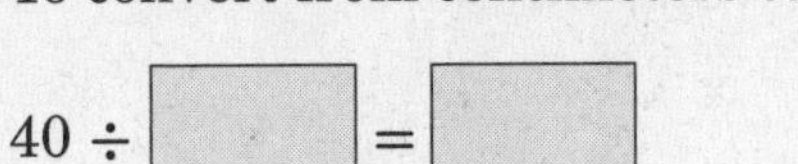

40 ÷ ____ = ____

So, 40 cm = ____ m.

FOLDABLES

ORGANIZE IT

Under the metric units tab, take notes on how to change metric units, include examples involving length, capacity, and mass.

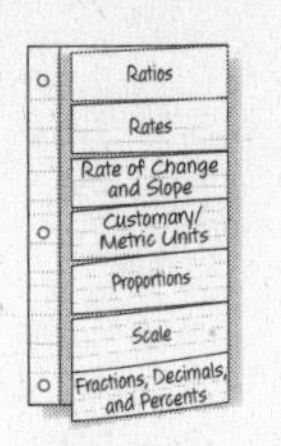

Check Your Progress **Complete.**

a. 7.5 m = ■ cm

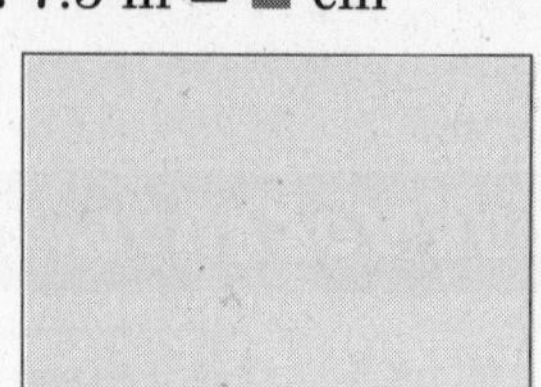

b. 3,400 mm = ■ m

EXAMPLE

WRITE IT

Explain how you can multiply a number by a power of ten.

______

______

______

______

3 **FARMS** **A bucket holds 12.8 liters of water. Find the capacity of the bucket in milliliters.**

You are converting from ______ to milliliters. Since the bucket holds 12.8 liters, use the relationship 1 L = ______ mL.

1 L = 1,000 mL	Write the relationship.
______ × 1 L = 12.8 × 1,000 mL	Multiply each side by 12.8 since you have 12.8 liters.
12.8 L = ______ mL	To multiply 12.8 by 1,000, move the decimal point ______ places to the right.

So, the capacity of the bucket in milliliters is  mL.

Check Your Progress **BOOKS** A box of textbooks has a mass of 32,850 grams. What is the mass of the box in kilograms?

HOMEWORK ASSIGNMENT

Page(s):

Exercises:

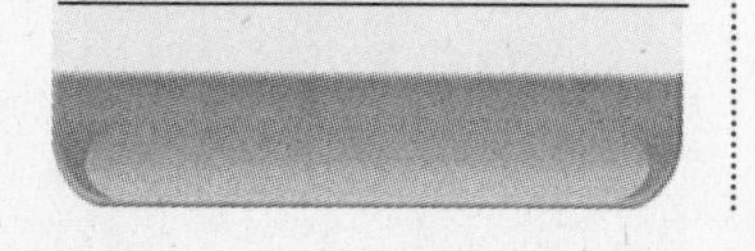

EXAMPLES Convert Between Measurement Systems

4 Convert 7.13 miles to kilometers. Round to the nearest hundredth if necessary.

Use the relationship 1 ____ ≈ 1.61 kilometers.

1 mi ≈ ____ km — Write the relationship.

7.13 × 1 mi ≈ 7.13 × ____ km — Multiply each side by ____ since you have 7.13 mi.

7.13 mi ≈ ____ km — Simplify.

So, 7.13 miles is approximately ____ kilometers.

5 Convert 925.48 grams to pounds. Round to the nearest hundredth if necessary.

Since 1 pound ≈ ____ grams, the unit ratio is $\frac{1 \text{ lb}}{453.6 \text{ g}}$.

925.48g ≈ ____ g · $\frac{1 \text{ lb}}{453.6 \text{ g}}$. — Multiply by ____.

≈ $\frac{925.48 \text{ lb}}{453.6}$ or ____ lb — Simplify.

So, 925.48 grams is approximately ____ pounds.

Check Your Progress **Complete. Round to the nearest hundredth if necessary.**

a. 8.15 gal = ■ L

b. 5.75 m = ■ yd

6-6 Algebra: Solving Proportions

MAIN IDEA

- Solve proportions.

BUILD YOUR VOCABULARY (pages 121–122)

Two quantities are **proportional** if they have a ______ rate or ratio.

A **proportion** is an equation stating that two ratios or rates are ______.

In a proportion, a **cross product** is the ______ of the numerator of one ratio and the denominator of the other ratio.

KEY CONCEPT

Proportion A proportion is an equation stating that two ratios are equivalent.

EXAMPLE Identify Proportional Relationships

1 MATH **Before dinner, Mohammed solved 8 math problems in 12 minutes. After dinner, he solved 2 problems in 3 minutes. Is the number of problems he solved proportional to the time?**

To identify proportional relationships, you can compare unit rates or compare ratios by comparing cross products. Let's compare ratios by comparing ______.

problems → $\frac{8}{12} \stackrel{?}{=} \frac{2}{3}$ ← problems
minutes → ← minutes

$8 \cdot 3 = \square \cdot 2$

$24 = 24$

Since the cross products are ______, the number of problems solved is proportional to the time.

Check Your Progress Determine if the quantities \$30 for 12 gallons of gasoline and \$10 for 4 gallons of gasoline are proportional.

EXAMPLES Solve a Proportion

2 **Solve $\frac{5}{8} = \frac{18}{x}$.**

$\frac{5}{8} = \frac{18}{x}$	Write the proportion.
$5 \cdot x = 8 \cdot 18$	Find the cross products.
$5x =$ ____	Multiply.
$\frac{5x}{\_\_\_\_} = \frac{144}{\_\_\_\_}$	Divide each side by ____.
$x =$ ____	Simplify.

FOLDABLES

ORGANIZE IT

Under the proportions tab, take notes on how to solve a proportion. Include examples.

3 **Solve $\frac{3.5}{14} = \frac{6}{n}$.**

$\frac{3.5}{14} = \frac{6}{n}$	Write the proportion.
$3.5 \cdot n = 14 \cdot 6$	Find the cross products.
$3.5n =$ ____	Multiply.
$\frac{3.5n}{\_\_\_\_} = \frac{84}{\_\_\_\_}$	Divide each side by ____.
$n =$ ____	Simplify.

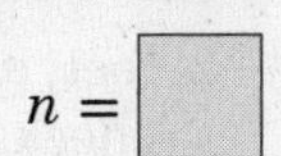

HOMEWORK ASSIGNMENT

Page(s):

Exercises:

Check Your Progress **Solve each proportion.**

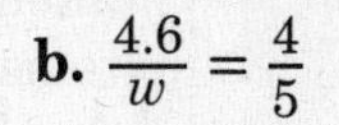

a. $\frac{9}{15} = \frac{k}{18}$ **b.** $\frac{4.6}{w} = \frac{4}{5}$

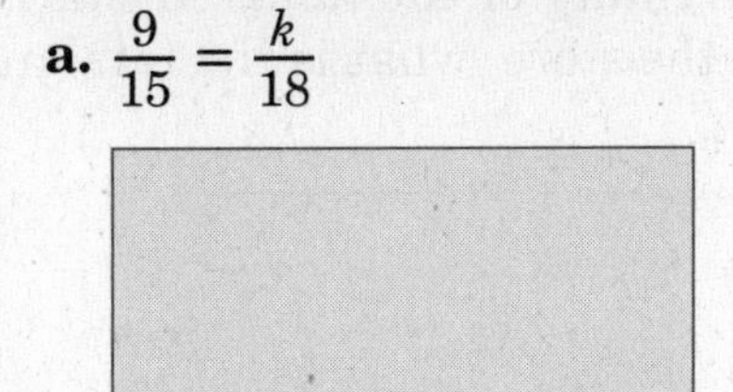

6–7 Problem-Solving Investigation: Draw a Diagram

EXAMPLE Draw a Diagram

MAIN IDEA

- Solve problems by drawing a diagram.

ROCK CLIMBING **A rock climber stops to rest at a ledge 90 feet above the ground. If this represents 75% of the total climb, how high above the ground is the top of the rock?**

UNDERSTAND You know that ______ feet is 75% of the total height. You need to find the total height.

PLAN Draw a diagram showing the part already climbed.

SOLVE

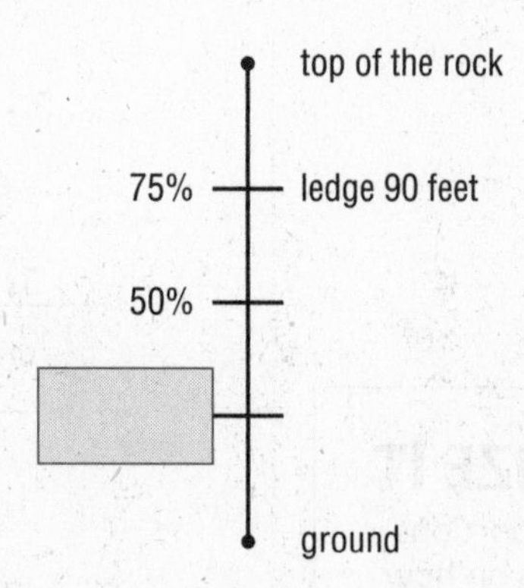

You know that 75% ÷ 3 = 25%. If 75% of the total height is 90 feet, then 25% of the total height would be 90 ÷ 3, or 30, feet. You know that 75% + 25% = ______, so 90 feet + 30 feet = 120 feet, which is the height of the top of the rock.

CHECK Since 75%, or 0.75, of the total height is 90 feet, and 90 ÷ 120 = ______, the solution checks.

Check Your Progress **INVENTORY** A retail store has taken inventory of 400 items. If this represents 80% of the total items in the store, what is the total number of items in the store?

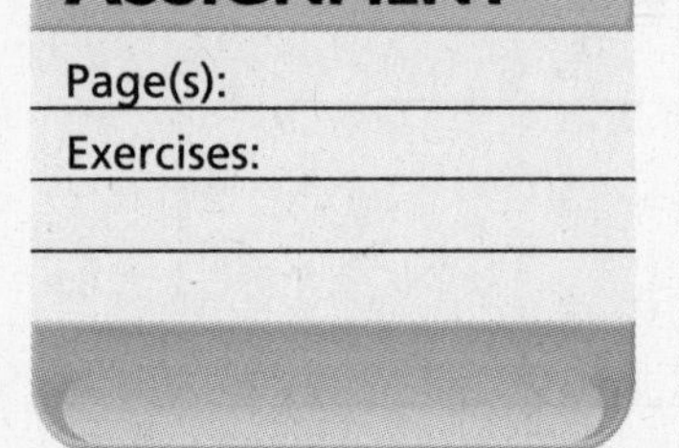

HOMEWORK ASSIGNMENT

Page(s):

Exercises:

6–8 Scale Drawings

Main Idea

- Solve problems involving scale drawings.

BUILD YOUR VOCABULARY (pages 121–122)

Scale drawings and **scale models** are used to represent objects that are too ______ or too ______ to be drawn at actual size.

The **scale** gives the ratio that compares the ______ of the drawing to the real object.

FOLDABLES

ORGANIZE IT

Under the scale tab, explain how to solve a problem involving scale drawings. Be sure to include an example.

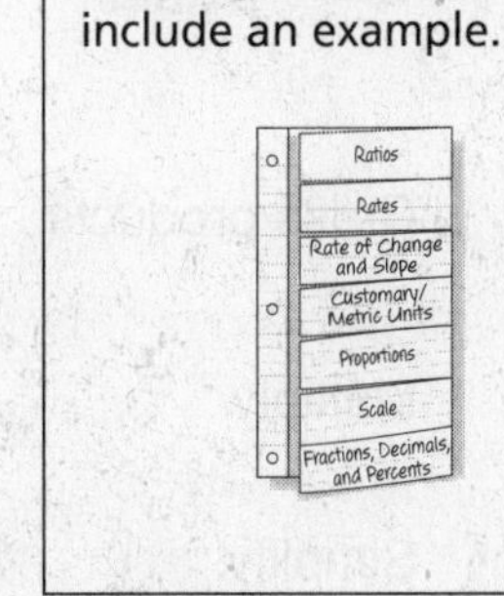

EXAMPLE Use a Map Scale

1 MAPS **What is the actual distance between Portland and Olympia?**

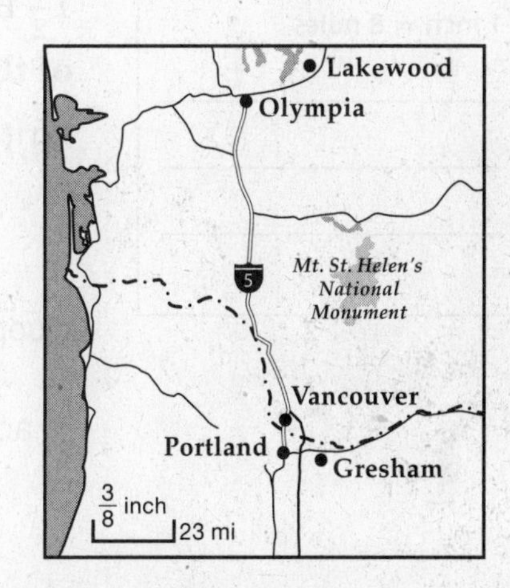

Step 1 Use a ruler to find the map distance between the two cities. The map distance is about ______.

Step 2 Write and solve a proportion using the scale. Let d represent the actual distance between the cities.

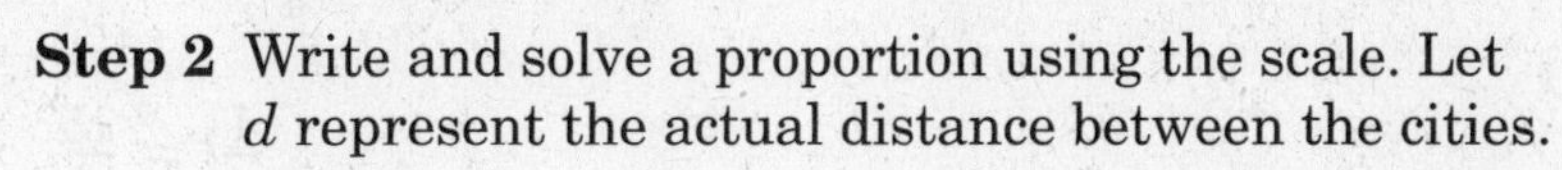

map → $\frac{\frac{3}{8}\text{ inch}}{23\text{ mi}} = \frac{1.69\text{ inches}}{d\text{ mi}}$ ← map

actual → ← actual

$\frac{3}{8} \times d = 23 \times 1.69$ Cross products.

$0.375d = 3.887$ Multiply. Write $\frac{3}{8}$ as a decimal.

$d =$ ______ Divide both sides by 0.375.

The distance between the cities is about ______ kilometers.

Check Your Progress **MAPS** On a map of California, the distance between San Diego and Bakersfield is about $11\frac{2}{5}$ centimeters. What is the actual distance if the scale is 1 centimeter = 30 kilometers?

WRITE IT

Explain why these two scales are equivalent scales:

$\frac{1}{2}$ inch = 4 miles

1 inch = 8 miles

EXAMPLE Use a Blueprint Scale

2 **ARCHITECTURE On the blueprint of a new house, each square has a side length of $\frac{1}{4}$ inch. If the length of a bedroom on the blueprint is $1\frac{1}{2}$ inches, what is the actual length of the room?**

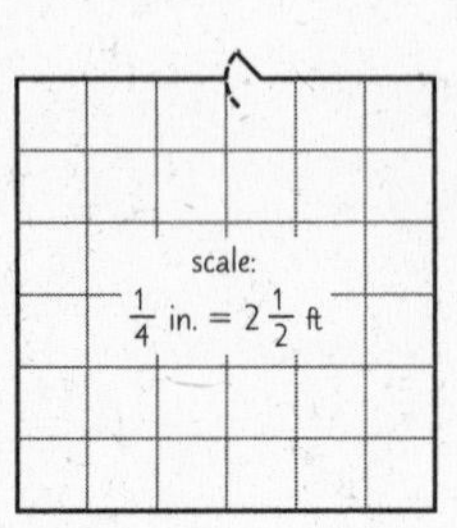

Write and solve a proportion.

	Scale		Length of Room	
blueprint →	$\frac{1}{4}$ inch	=	___	← blueprint
actual →	___		t feet	← actual

$\frac{1}{4} \cdot t =$ ___ Cross products

$\frac{1}{4}t = \frac{15}{4}$ Multiply.

$t =$ ___ Simplify.

The length of the room is ___.

Check Your Progress On a blueprint of a new house, each square has a side length of $\frac{1}{4}$ inch. If the width of the kitchen on the blueprint is 2 inches, what is the actual width of the room?

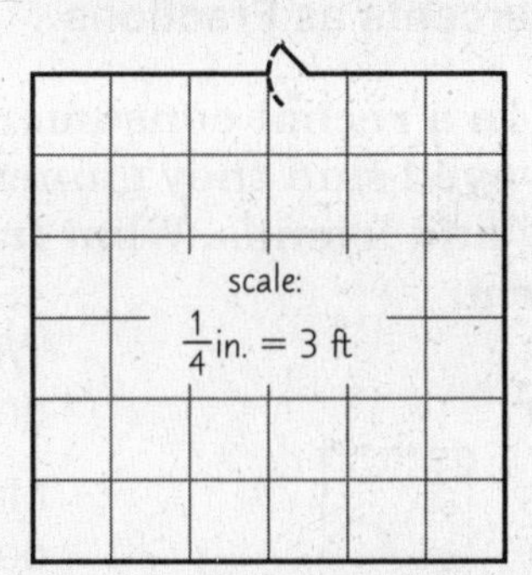

EXAMPLE Find a Scale Factor

3 Find the scale factor of a blueprint if the scale is $\frac{1}{2}$ inch = 3 feet.

$\frac{\frac{1}{2}\text{ inch}}{3\text{ feet}} = \frac{\frac{1}{2}\text{ inch}}{\square}$ Convert 3 feet to ______.

$= \square \cdot \frac{\frac{1}{2}\text{ inch}}{36\text{ inches}}$ Multiply by ______ to eliminate the fraction in the numerator.

$= \square$ Divide out the common units.

The scale factor is ______. That is, each measure on the blueprint is ______ the ______ measure.

HOMEWORK ASSIGNMENT

Page(s):

Exercises:

Check Your Progress Find the scale factor of a blueprint if the scale is 1 inch = 4 feet.

6–9 Fractions, Decimals, and Percents

MAIN IDEA

- Write percents as fractions and decimals and vice versa.

EXAMPLES Percents as Fractions

1 **NUTRITION** **In a recent consumer poll, 41.8% of the people surveyed said they gained nutrition knowledge from family and friends. What fraction is this? Write in simplest form.**

$41.8\% = \frac{41.8}{100}$ Write a fraction with a denominator of 100.

$= \frac{41.8}{100} \cdot$ ____ Multiply to eliminate the decimal in the numerator.

$=$ ____ or ____ Simplify.

2 **Write $12\frac{1}{2}\%$ as a fraction in simplest form.**

$12\frac{1}{2}\% = \frac{12\frac{1}{2}}{100}$ Write a fraction.

$= 12\frac{1}{2} \div 100$ Divide.

$=$ ____ $\div 100$ Write $12\frac{1}{2}$ as an improper fraction.

$=$ ____ $\times$ ____ Multiply by the reciprocal of 100.

$=$ ____ or ____ Simplify.

FOLDABLES

ORGANIZE IT

Under the Fractions, Decimals, and Percents tab, take notes on writing percents as fractions and fractions as percents. Include examples.

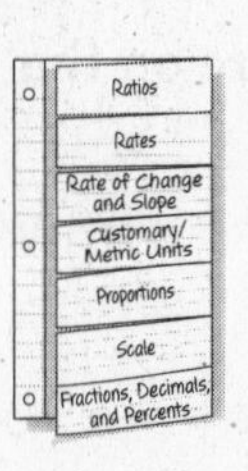

Check Your Progress

a. **ELECTION** In a recent election, 64.8% of registered voters actually voted. What fraction is this? Write in simplest form.

b. Write $62\frac{1}{2}\%$ as a fraction in simplest form.

KEY CONCEPTS

Common Fraction/ Decimal/Percent Equivalents

$\frac{1}{3} = 0.\overline{3} = 33\frac{1}{3}\%$

$\frac{2}{3} = 0.\overline{6} = 66\frac{2}{3}\%$

$\frac{1}{8} = 0.125 = 12\frac{1}{2}\%$

$\frac{3}{8} = 0.375 = 37\frac{1}{2}\%$

$\frac{5}{8} = 0.625 = 62\frac{1}{2}\%$

$\frac{7}{8} = 0.875 = 87\frac{1}{2}\%$

EXAMPLES Fractions as Percents

3 PRODUCE In one shipment of fruit to a grocery store, 5 out of 8 bananas were still green. Find this amount as a percent.

$\frac{5}{8} = \frac{n}{100}$ Write a proportion.

$500 = 8n$ Find the cross products.

$\frac{500}{\square} = \frac{8n}{\square}$ Divide each side by $\square$.

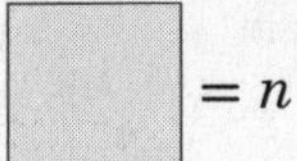

$\square = n$ Simplify.

So, $\frac{5}{8} = 62\frac{1}{2}\%$ or $\square$.

4 Write $\frac{5}{12}$ as a percent. Round to the nearest hundredth if necessary.

$\frac{5}{12} = \frac{n}{100}$ Write a proportion.

$\square = \square$ Find the cross products.

500 [÷] 12 [ENTER] 41.66666667 Use a calculator.

So, $\frac{5}{12}$ is about $\square$.

5 Write $\frac{3}{7}$ as a percent. Round to the nearest hundredth.

$\frac{3}{7} = 0.4285714...$ Write $\frac{3}{7}$ as a decimal.

$= \square$ $\square$ by 100 and add the $\square$.

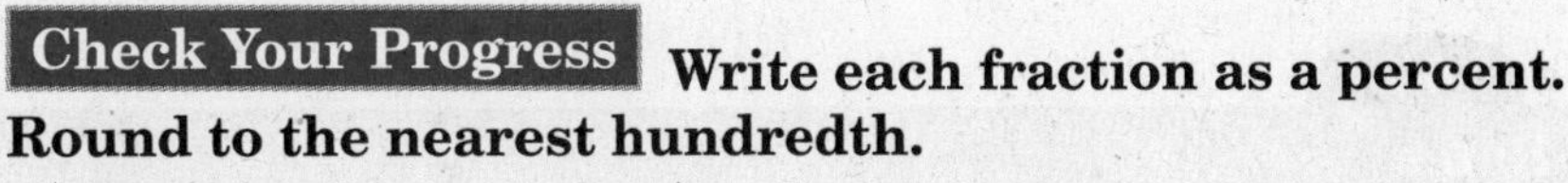

Check Your Progress Write each fraction as a percent. Round to the nearest hundredth.

a. $\frac{13}{25}$

HOMEWORK ASSIGNMENT

Page(s):

Exercises:

b. $\frac{11}{15}$

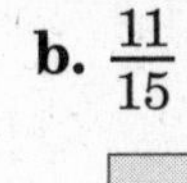

CHAPTER 6

BRINGING IT ALL TOGETHER

STUDY GUIDE

FOLDABLES	VOCABULARY PUZZLEMAKER	BUILD YOUR VOCABULARY
Use your **Chapter 6 Foldable** to help you study for your chapter test.	To make a crossword puzzle, word search, or jumble puzzle of the vocabulary words in Chapter 6, go to: glencoe.com	You can use your completed **Vocabulary Builder** (*pages 121–122*) to help you solve the puzzle.

6-1

Ratios

State whether each sentence is true or false. If false, replace the underlined word to make it a true sentence.

1. When you simplify a ratio, write a fraction as a mixed number.

2. To write a ratio comparing measures, both quantities should have the same unit of measure.

Write each ratio as a fraction in simplest form.

3. 63:7

4. 15:54

6-2

Rates

Complete.

5. A ______ is a ratio that compares two quantities with different kinds of units.

Write each ratio as a fraction in simplest form.

6. 36 inches: 48 inches

7. 15 minutes to 3 hours

6-3 Rate of Change and Slope

8. The table shows Amanda's running time during a 5-mile race. Graph the data. Find the slope of the line. Explain what the slope represents.

Distance (miles)	Time (minutes)
1	6
2	12
3	18
4	24
5	30

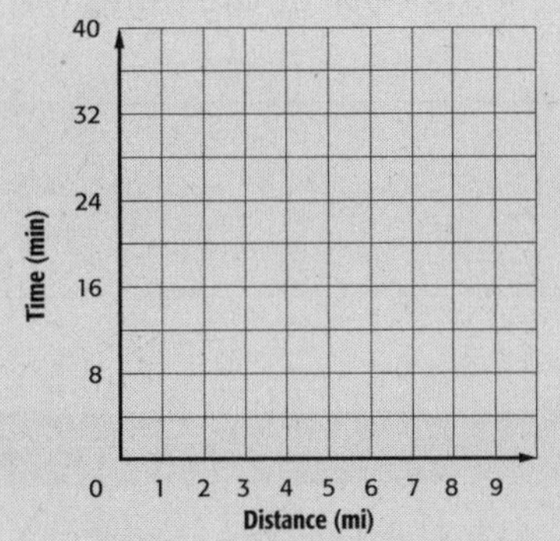

6-4 Measurement: Changing Customary Units

Complete.

9. $3\frac{3}{4}$ pt = ■ c

10. 90 ft = ■ yd

11. 156 oz = ■ lb

6-5 Measurement: Changing Metric Units

Complete.

12. 4.3 cm = ______ mm

13. 42.7 g = ______ mg

6-6 Algebra: Solving Proportions

Complete each sentence.

14. The cross products of a ______ are equal.

15. If you know ______ parts of a proportion, you can solve for the fourth part by ______ and then ______ both sides by the coefficient of the unknown.

Solve each proportion.

16. $\frac{15}{n} = \frac{3}{8}$ ______

17. $\frac{6}{20} = \frac{x}{80}$ ______

18. $\frac{b}{16} = \frac{3}{48}$ ______

6-7

Problem-Solving Investigation: Draw a Diagram

19. **LADDERS** A ladder leans against a wall. The top of the ladder rests against the wall at a point 12 feet above the ground. If this distance represents 80% of the height of the wall, how tall is the wall?

Scale Drawings

On a map, the scale is $\frac{1}{4}$ inch = 10 miles. For each map distance, find the actual distance.

20. 6 inches

21. $\frac{3}{8}$ inch

22. $2\frac{1}{2}$ inches

23. 1 inch

6-9

Fractions, Decimals, and Percents

Complete the table of equivalent fractions.

	Fraction	Decimal	Percent
24.	$\frac{1}{3}$		
25.	$\frac{3}{8}$		$37\frac{1}{2}\%$
26.	$\frac{1}{8}$		
27.		0.875	$87\frac{1}{2}\%$

ARE YOU READY FOR THE CHAPTER TEST?

Math Online

Visit **glencoe.com** to access your textbook, more examples, self-check quizzes, and practice tests to help you study the concepts in Chapter 6.

Check the one that applies. Suggestions to help you study are given with each item.

☐ **I completed the review of all or most lessons without using my notes or asking for help.**

- You are probably ready for the Chapter Test.
- You may want to take the Chapter 6 Practice Test on page 337 of your textbook as a final check.

☐ **I used my Foldable or Study Notebook to complete the review of all or most lessons.**

- You should complete the Chapter 6 Study Guide and Review on pages 333–336 of your textbook.
- If you are unsure of any concepts or skills, refer back to the specific lesson(s).
- You may also want to take the Chapter 6 Practice Test on page 337 of your textbook.

☐ **I asked for help from someone else to complete the review of all or most lessons.**

- You should review the examples and concepts in your Study Notebook and Chapter 6 Foldable.
- Then complete the Chapter 6 Study Guide and Review on pages 333–336 of your textbook.
- If you are unsure of any concepts or skills, refer back to the specific lesson(s).
- You may also want to take the Chapter 6 Practice Test on page 337 of your textbook.

Student Signature

Parent/Guardian Signature

Teacher Signature

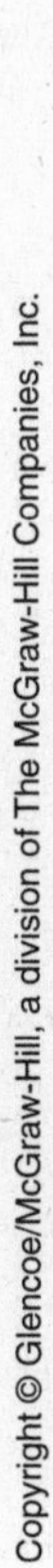

Applying Percents

Use the instructions below to make a Foldable to help you organize your notes as you study the chapter. You will see Foldable reminders in the margin of this Interactive Study Notebook to help you in taking notes.

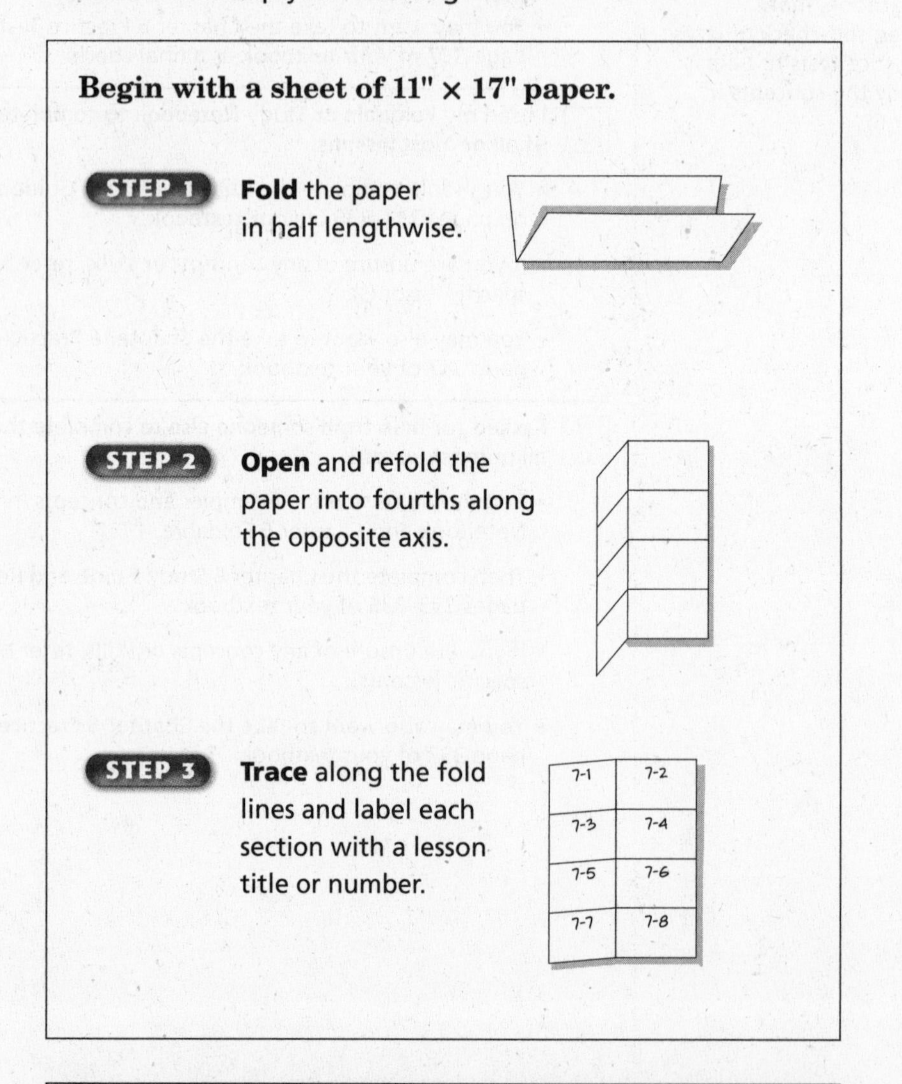

Begin with a sheet of 11" × 17" paper.

STEP 1 **Fold** the paper in half lengthwise.

STEP 2 **Open** and refold the paper into fourths along the opposite axis.

STEP 3 **Trace** along the fold lines and label each section with a lesson title or number.

NOTE-TAKING TIP: When you take notes, it is often helpful to reflect on ways the concepts apply to your daily life.

BUILD YOUR VOCABULARY

This is an alphabetical list of new vocabulary terms you will learn in Chapter 7. As you complete the study notes for the chapter, you will see Build Your Vocabulary reminders to complete each term's definition or description on these pages. Remember to add the textbook page number in the second column for reference when you study.

Vocabulary Term	Found on Page	Definition	Description or Example
discount			
percent equation			
percent of change			
percent of decrease			
percent of increase			

(continued on the next page)

Vocabulary Term	Found on Page	Definition	Description or Example
percent proportion			
principal			
sales tax			
simple interest			

7-1 Percent of a Number

MAIN IDEA

- Find the percent of a number.

EXAMPLE Find the Percent of a Number

1 Find 8% of 125.

METHOD 1 Write the percent as a fraction.

$8\% = \frac{8}{100}$ or ______

$\frac{2}{25}$ of $125 = \frac{2}{25} \times 125$ or ______

METHOD 2 Write the percent as a decimal.

$8\% = \frac{8}{100}$ or ______

0.08 of $125 = 0.08 \times 125$ or ______

So, 8% of 125 is ______.

REMEMBER IT

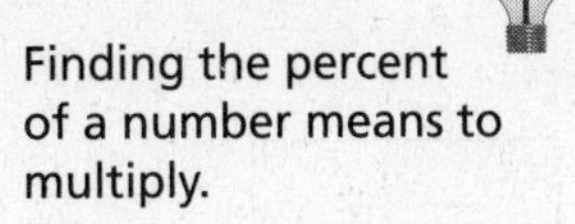

Finding the percent of a number means to multiply.

Check Your Progress Find 72% of 350.

EXAMPLE Use Percents Greater than 100%

2 Find 125% of 64.

You can either write the percent as a ______ or as a ______. Let's write the percent as a decimal.

$125\% = \frac{125}{100} =$ ______

1.25 of $64 = 1.25 \times 64$ or ______

So, 125% of 64 is ______.

Check Your Progress Find 225% of 50.

EXAMPLE

3 LANGUAGES The graph below shows that 30% of the people in a community speak Spanish as their first language. If a community has 800 people, how many people can be expected to speak Spanish as their first language?

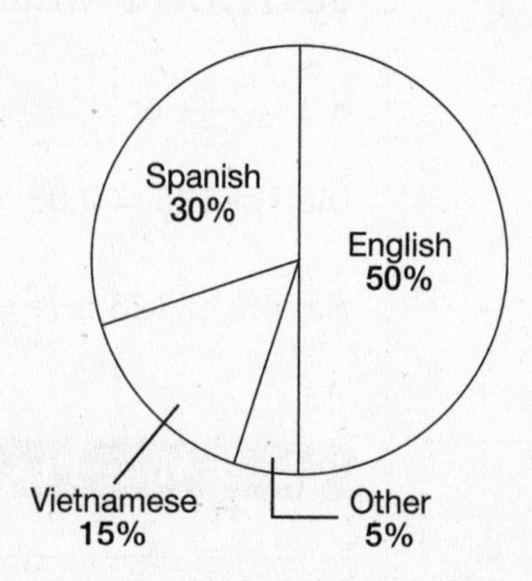

To find 30% of 800, write the percent as a ______. Then multiply.

30% of $800 = 30\% \cdot 800$

$= \_\_\_\_ \cdot 800$

$= 240$

So, about ____ people in the community speak Spanish as their first language.

Check Your Progress **SLEEP** The average person sleeps 33% of their adult life. If their adult life consists of 62 years, how many years does the average person spend sleeping?

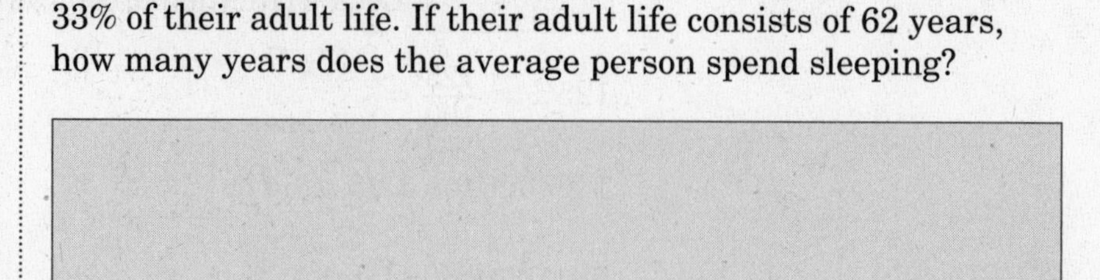

HOMEWORK ASSIGNMENT

Page(s):

Exercises:

7-2 The Percent Proportion

MAIN IDEA

- Solve problems using the percent proportion.

BUILD YOUR VOCABULARY (pages 149–150)

A **percent proportion** compares ______ of a quantity to the whole quantity, called the ______, using a percent.

KEY CONCEPT

Percent Proportion The percent proportion is $\frac{\text{part}}{\text{whole}} = \frac{\text{percent}}{100}$.

EXAMPLE Find the Percent

1 **What percent of 24 is 18?**

18 is the part, and 24 is the whole. You need to find the percent.

$\frac{p}{w} = \frac{n}{100}$ Write the proportion.

______ $= \frac{n}{100}$ $p =$ ______, $w =$ ______

$18 \cdot 100 = 24 \cdot n$ Find the cross products.

$1{,}800 = 24n$ Simplify.

______ $= \frac{24n}{24}$ Divide each side by ______.

______ $= n$ Simplify.

So, ______ of 24 is ______.

EXAMPLE Find the Part

2 **What number is 30% of 150?**

30 is the percent and 150 is the base. You need to find the part.

$\frac{p}{w} = \frac{n}{100}$ Percent proportion

$\frac{p}{150} =$ ______ $w =$ ______, $n =$ ______

$p \cdot 100 = 150 \cdot 30$ Find the cross products.

$100p =$ ______ Simplify.

$\frac{100p}{100} = \frac{4{,}500}{100}$ Divide each side by 100.

$p =$ ______ Simplify.

So, 30% of ______ is 45.

EXAMPLE Find the Base

3 12 is 80% of what number?

12 is the part and 80 is the percent. You need to find the base.

$\frac{p}{w} = \frac{n}{100}$ Percent proportion

$\frac{12}{w} =$ ______ $a =$ ______ , $n = 80$.

______ $= w \cdot 80$ Find the cross products.

$1{,}200 =$ ______ Simplify.

$\frac{1{,}200}{8} = \frac{80w}{80}$ Divide each side by ______.

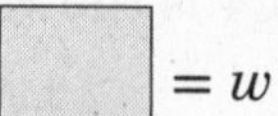
$= w$

So, 12 is 80% of 15.

WRITE IT

Write an example of a real-world percent problem.

Check Your Progress

a. What percent of 80 is 28?

b. What number is 65% of 180?

c. 36 is 40% of what number?

HOMEWORK ASSIGNMENT

Page(s):

Exercises:

7–3 Percent and Estimation

MAIN IDEA

- Estimate percents by using fractions and decimals.

EXAMPLE

1 CONCERTS A town sold 407 tickets to a chamber music concert in the town square. Of the tickets sold, 61% were discounted for senior citizens. About how many senior citizens bought tickets for the concert?

You need to estimate 61% of 407.

61% is about 60%, and 407 is about 400.

61% of 407 ≈ ______ · 400 $61\% \approx \frac{3}{5}$

≈ 240 Multiply.

So, about ______ senior citizens bought tickets.

Check Your Progress **TAXES** Michelle discovered that 27% of her paycheck was deducted for taxes. If her paycheck before taxes was $590, about how much was deducted for taxes?

EXAMPLE

2 COINS Melinda calculated that 40% of the coins in her coin collection were minted before 1964. If there are 715 coins in her collection, about how many of them were minted before 1964?

You can use a fraction or 10% of a number to estimate. Let's use 10% of a number.

Step 1 Find 10% of the number.

715 is about ______.

10% of 700 = 0.1 · 700

= ______

FOLDABLES

ORGANIZE IT

Record the main ideas, and give examples about percent and estimation in the section for Lesson 7-3 of your Foldable.

7-1	7-2
7-3	7-4
7-5	7-6
7-7	7-8

(continued on the next page)

Step 2 Multiply.

40% of 700 is 4 • 10% of 700.

$4 \times 70 =$ ______

So, about ______ coins were minted before 1964.

Check Your Progress **SAVINGS** Suki saves 70% of her monthly allowance. If her monthly allowance is $58, about how much does she save?

REMEMBER IT

To estimate the percent of a number, round the percent, round the number, or round both.

EXAMPLES Percents Greater Than 100 or Less Than 1

3 **Estimate 173% of 60.**

173% is about 175%.

$175\% \text{ of } 60 = (100\% \text{ of } 60) + (75\% \text{ of } 60)$

$= (1 \cdot 60) + \left(\frac{3}{4} \cdot 60\right)$

$= 60 + 45$ or ______

So, 173% of 60 is about ______.

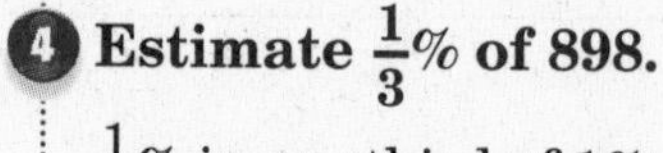

4 **Estimate $\frac{1}{3}$% of 898.**

$\frac{1}{3}$% is one third of 1%. 898 is about 900.

$1\% \text{ of } 900 = 0.01 \cdot 900$ Write 1% as ______.

$= 9$ Multiply.

One third of 9 is $\frac{1}{3} \cdot 9$ or ______.

So, $\frac{1}{3}$% of 898 is about ______.

HOMEWORK ASSIGNMENT

Page(s):

Exercises:

Check Your Progress **Estimate.**

a. 142% of 80

b. $\frac{1}{5}$% of 197

7–4 Algebra: The Percent Equation

MAIN IDEA

- Solve problems by using the percent equation.

BUILD YOUR VOCABULARY (pages 149–150)

The equation ______ = percent • ______ is called the **percent equation.**

FOLDABLES

ORGANIZE IT

Record the main ideas, and give examples about the percent equation in the section for Lesson 7-4 of your Foldable.

7-1	7-2
7-3	7-4
7-5	7-6
7-7	7-8

EXAMPLE Find the Part

1 What number is 46% of 200?

46% or ______ is the percent and ______ is the whole.

Let p represent the ______.

part = percent • whole

p = ______ • 200 Write an equation.

p = ______ Multiply.

So, 46% of 200 is ______.

EXAMPLE Find the Percent

2 26 is what percent of 32?

Let n represent the percent.

part = percent • whole

______ = n • 32 Write an equation.

______ = ______ Divide each side by ______.

______ = n Simplify.

______ = n Write as a percent.

So, 26 is ______ of 32.

EXAMPLE Find the Whole

3 12 is 40% of what number?

Let w represent the whole.

part = percent · whole

$\square = \square \cdot w$ Write an equation.

$\frac{\square}{0.40} = \frac{w}{\square}$ Divide each side by $\square$.

$\square = w$

So, 12 is 40% of $\square$.

Check Your Progress

a. What number is 72% of 500?

b. 18 is what percent of 80?

c. 36 is 90% of what number?

WRITE IT

Name two ways a percent can be written in the percent equation.

____

____

____

____

____

HOMEWORK ASSIGNMENT

Page(s):

Exercises:

7–5 Problem-Solving Investigation: Determine Reasonable Answers

EXAMPLE Solve. Use the Reasonable Answer Strategy.

MAIN IDEA

- Solve problems by determining reasonable answers.

FUNDRAISER A soccer team is having a candy sale to raise funds to buy new shirts. The team gets to keep 25% of the sales. Each candy bar costs $1.50, and the team has sold 510 bars so far. If the shirts cost a total of $175, should the team order the shirts yet? Explain.

UNDERSTAND You know the shirts cost a total of $175 and that each candy bar costs $1.50. You know that the team has sold ______ bars so far and that they get to keep 25% of the sales. You need to know if the team has enough money to order the shirts yet.

PLAN Find how much the team has earned so far. Round 510 to 500. Then find ______ of their sales.

SOLVE $1.50 • 500 = ______

Find 25% of $750.

25% of 750 = 0.25 • 750

= ______

The team gets to keep ______. Since this is more than the cost of the shirts, they should order the shirts.

CHECK Use a calculator to check that the actual result is 191.25, so the answer is reasonable.

Check Your Progress **FIELD TRIP** There are 392 students in the seventh grade at Hamilton Middle School. If 35% of the seventh grade will attend the class field trip, is it reasonable to say that about 170 students will attend the field trip? Explain.

HOMEWORK ASSIGNMENT

Page(s):

Exercises:

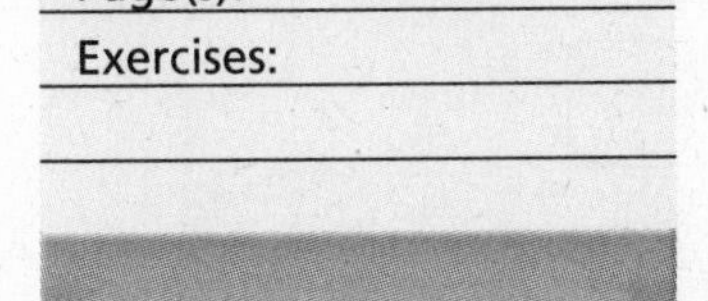

7–6 Percent of Change

MAIN IDEA

- Find the percent of increase or decrease.

BUILD YOUR VOCABULARY (pages 149–150)

A **percent of change** is a ratio that compares the change in quantity to the ______ amount.

If the **original** quantity is ______, the percent of change is called the **percent of increase**.

If the **original** quantity is ______, the percent of change is called the **percent of decrease**.

EXAMPLE Find Percent of Increase

1 SHOPPING **Last year a sweater sold for \$56. This year the same sweater sells for \$60. Find the percent of change in the cost of the sweater. Round to the nearest whole percent if necessary.**

Since the new price is ______ than the original price, this is a percent of ______. The amount of increase is 60 – ______ or ______.

percent of increase = $\dfrac{\text{amount of increase}}{\text{\_\_\_\_\_\_}}$

= $\dfrac{\text{\_\_\_\_\_\_}}{56}$ Substitution

= ______ Simplify.

= ______ Write as a ______.

The percent of ______ is about ______.

Check Your Progress **DVDs** Last year a DVD sold for $20. This year the same DVD sells for $24. Find the percent of change in the cost of the DVD. Round to the nearest whole percent if necessary.

____

EXAMPLE Find Percent of Decrease

FOLDABLES

ORGANIZE IT

Record the main ideas, and give examples about percent of change in the section for Lesson 7-6 of your Foldable.

2 ATTENDANCE **On the first day of school this year, 435 students reported to Howard Middle School. Last year on the first day, 460 students attended. Find the percent of change for the first day attendance. Round to the nearest whole percent if necessary.**

Since the new enrollment figure is ____ than the figure for ____ year, this is a percent of ____. The amount of decrease is ____ – 435 or ____ students.

percent of decrease = $\frac{\text{\_\_\_\_}}{\text{original amount}}$

= $\frac{25}{\text{\_\_\_\_}}$ Substitution

= ____ Simplify.

= ____ Write ____ as a percent.

The percent of ____ in the enrollment is about ____.

Check Your Progress **ZOO** At the beginning of the summer season, the local zoo reported having 385 animals in its care. At the beginning of last year's summer season the zoo had reported 400 animals. Find the percent of change in the number of animals at the zoo. Round to the nearest whole percent if necessary.

____

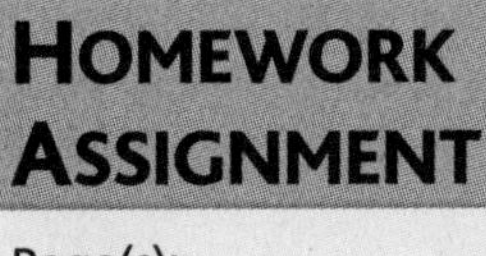

Page(s):

Exercises:

7–7 Sales Tax and Discount

Main Idea

- Solve problems involving sales tax and discount.

BUILD YOUR VOCABULARY (pages 149–150)

Sales tax is an ______ amount of money charged on items that people ______.

Discount is the amount by which the regular ______ of an item is ______.

EXAMPLE Find the Total Cost

1 GOLF A set of golf balls sells for $20, and the sales tax is 5.75%. What is the total cost?

To find the total cost, you can add sales tax to the regular price or add the percent of tax to 100%. Let's add sales tax to the regular price.

First, find the ______ tax.

5.75% of $20 = ______ • 20

= ______ The sales tax is ______.

Next, add the sales tax to the regular price.

______ + 20 = ______

The ______ cost of the set of golf balls is ______.

FOLDABLES ORGANIZE IT

Record the main ideas, and give examples about sales tax and discount in the section for Lesson 7-7 of your Foldable.

7-1	7-2
7-3	7-4
7-5	7-6
7-7	7-8

Check Your Progress **BOOKS** A set of three paperback books sells for $35 and the sales tax is 7%. What is the total cost of the set?

EXAMPLE Find the Sale Price

REMEMBER IT

The cost of an item with sales tax will always be greater than the regular price. The discounted price of an item is always less than the regular price.

2 OUTERWEAR **Whitney wants to buy a new coat that has a regular price of $185. This weekend, the coat is on sale at a 33% discount. What is the sale price of the coat?**

METHOD 1

First, find the amount of the ______ *d*.

33% of $185 = ______ • $185 Write 33% as a decimal.

= ______ The discount is $61.05.

So, the sale price is $185 − ______ or ______.

METHOD 2

First, subtract the ______ of discount from 100%.

100% − ______ = ______

So, the sale price is ______ of the regular price.

67% of $185 = ______ • 185 Write 67% as a decimal.

= ______ Use a calculator.

So, the sale price of the coat is ______.

Check Your Progress **ELECTRONICS** Alex wants to buy a DVD player that has a regular price of $175. This weekend, the DVD player is on sale at a 20% discount. What is the sale price of the DVD player?

EXAMPLE Find the Percent of the Discount

3 WATCHES **A sports watch is on sale for $60.20 after a 30% discount. What is the original price?**

First, find the percent paid.

100% – 30% = ______

Next, use the ______ equation to find the ______.

Words	______ is 70% of what amount?
Variable	Let n represent the original price.
Equation	$60.20 = 70\% \cdot$ ______

$60.20 =$ ______ $\cdot\, n$ Write 70% as a decimal.

______ $= n$ ______ each side by 0.70.

The original price of the sports watch is ______.

Check Your Progress **FURNITURE** A rocking chair is on sale for $318.75 after a 15% discount. What is the original price?

Page(s):

Exercises:

7-8 Simple Interest

BUILD YOUR VOCABULARY (pages 149–150)

Simple Interest is the amount ______ or earned for the use of money.

Principal is the amount of ______ deposited or ______.

MAIN IDEA

- Solve problems involving simple interest.

EXAMPLES Find Interest Earned

SAVINGS Brandon found a bank offering a certificate of deposit that pays 4% simple interest. He has $1,500 to invest. How much interest will he earn in each amount of time?

1 3 years

$I = prt$ Formula for simple interest

$I =$ ______ · ______ · ______ Replace the variables.

$I =$ ______ Simplify.

Brandon will earn ______ in interest in ______ years.

FOLDABLES ORGANIZE IT

Record the main ideas, and give examples about simple interest in the section for Lesson 7-8 of your Foldable.

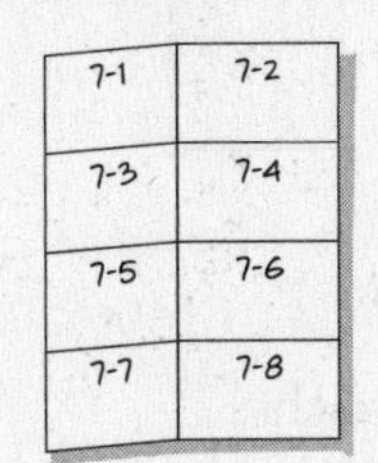

2 30 months

30 months = ______ = ______ years Write the time as years.

$I = prt$ Formula for simple interest

$I =$ ______ · ______ · ______ Replace the variables.

$I =$ ______ Simplify.

Brandon will earn  in interest in 30 months.

WRITE IT

Which is better: a higher percentage of interest on your credit card or on your savings account? Explain.

Check Your Progress

a. **SAVINGS** Cheryl opens a savings account that pays 5% simple interest. She deposits $600. How much interest will she earn in 2 years?

b. **SAVINGS** Micah opens a savings account that pays 4% simple interest. He deposits $2,000. How much interest will he earn in 42 months?

EXAMPLE Find Interest Paid on a Loan

3 **LOANS Laura borrowed $2,000 from her credit union to buy a computer. The interest rate is 9% per year. How much interest will she pay if it takes 8 months to repay the loan?**

$I =$ ______ Formula for simple interest

$I = 2{,}000 \cdot 0.09 \cdot \frac{8}{12}$ Replace p with ______, r with ______, and t with ______.

$I =$ ______ Simplify.

Laura will pay ______ in interest in ______ months.

Check Your Progress **LOANS** Juan borrowed $7,500 from the bank to purchase a used car. The interest rate is 15% per year. How much interest will he pay if it takes 2 years to repay the loan?

HOMEWORK ASSIGNMENT

Page(s):

Exercises:

CHAPTER 7

BRINGING IT ALL TOGETHER

STUDY GUIDE

FOLDABLES	VOCABULARY PUZZLEMAKER	BUILD YOUR VOCABULARY
Use your **Chapter 7 Foldable** to help you study for your chapter test.	To make a crossword puzzle, word search, or jumble puzzle of the vocabulary words in Chapter 7, go to: glencoe.com	You can use your completed **Vocabulary Builder** (*pages 149–150*) to help you solve the puzzle.

7-1 Percent of a Number

Find each number.

1. What is 3% of 530?

2. Find 15% of $24.

3. Find 200% of 17.

4. What is 0.6% of 800?

7-2 The Percent Proportion

5. In the formula $\frac{p}{w} = \frac{n}{100}$, p is the 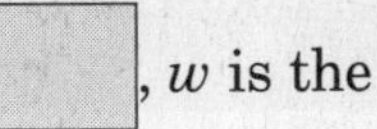, w is the ,

and n is the ____.

6. What number is 30% of 15?

7. 32.5 is 65% of what number?

7-3

Percent and Estimation

Write fraction equivalents in simplest form for the following percents.

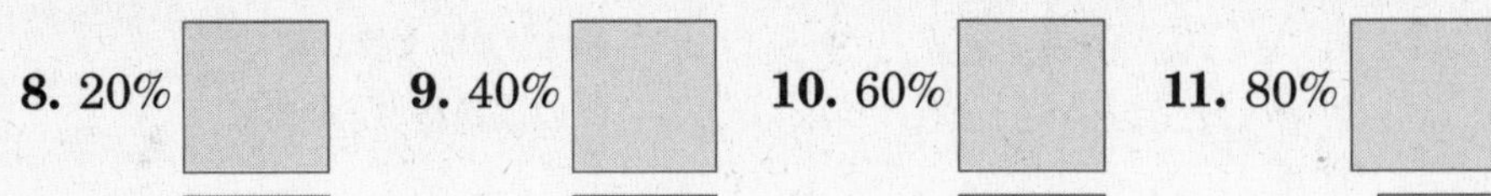

8. 20% **9.** 40% **10.** 60% **11.** 80%

12. 25% **13.** 50% **14.** 75% **15.** 100%

Estimate.

16. 49% of 80

17. 78% of 25

18. 153% of 10

19. 0.5% of 200

7-4

Algebra: The Percent Equation

Write an equation for each problem. Then solve.

20. 40% of what number is 48?

21. 18 is what percent of 72?

22. Find 80% of 90.

23. 12% of what number is 60?

7-5

Problem-Solving Investigation: Determine Reasonable Answers

24. TRAVEL The Winston family determined that lodging accounted for 48% of their total travel costs. If they spent $1,240 total during their trip, would about $560, $620, or $750 be a reasonable amount that they spent on lodging?

7-6

Percent of Change

State whether each sentence is *true* or *false*. If *false*, replace the underlined word to make a true sentence.

25. If the new amount is less than the original amount, then there is a percent of <u>increase</u>.

26. The amount of increase is the new amount <u>minus</u> the original amount.

Find the percent of change. Round to the nearest whole percent. State whether the percent of change is an increase or decrease.

27. original: $48; new: $44.25

28. original: $157; new: $181

29. original: $17.48; new: $9.98

7-7

Sales Tax and Discount

Find the total cost or sale price to the nearest cent.

30. $29.99 jeans; 15% discount

31. $6.25 lunch; 8.5% sales tax

Find the percent of discount to the nearest percent.

32. Pen: regular price, $9.95; sale price, $6.95

33. Sweatshirt: regular price, $20; sale price, $15.95

7-8

Simple Interest

Find the interest earned to the nearest cent for each principal, interest rate, and time.

34. $15,000, 9%, 2 years, 4 months

35. $250, 3.5%, 6 years

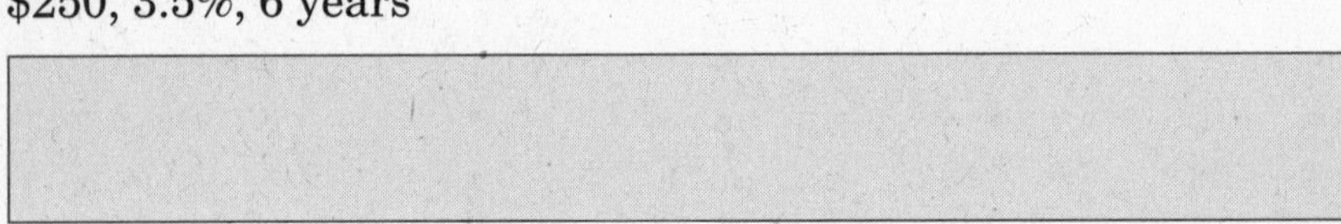

ARE YOU READY FOR THE CHAPTER TEST?

Math Online

Visit **glencoe.com** to access your textbook, more examples, self-check quizzes, and practice tests to help you study the concepts in Chapter 7.

Check the one that applies. Suggestions to help you study are given with each item.

- [ ] **I completed the review of all or most lessons without using my notes or asking for help.**
  - You are probably ready for the Chapter Test.
  - You may want to take the Chapter 7 Practice Test on page 389 of your textbook as a final check.

- [ ] **I used my Foldable or Study Notebook to complete the review of all or most lessons.**
  - You should complete the Chapter 7 Study Guide and Review on pages 384–388 of your textbook.
  - If you are unsure of any concepts or skills, refer back to the specific lesson(s).
  - You may also want to take the Chapter 7 Practice Test on page 389 of your textbook.

- [ ] **I asked for help from someone else to complete the review of all or most lessons.**
  - You should review the examples and concepts in your Study Notebook and Chapter 7 Foldable.
  - Then complete the Chapter 7 Study Guide and Review on pages 384–388 of your textbook.
  - If you are unsure of any concepts or skills, refer back to the specific lesson(s).
  - You may also want to take the Chapter 7 Practice Test on page 389 of your textbook.

Student Signature

Parent/Guardian Signature

Teacher Signature

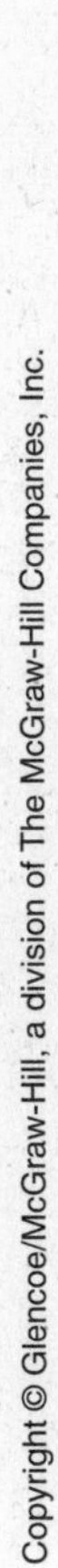

Statistics: Analyzing Data

Use the instructions below to make a Foldable to help you organize your notes as you study the chapter. You will see Foldable reminders in the margin of this Interactive Study Notebook to help you in taking notes.

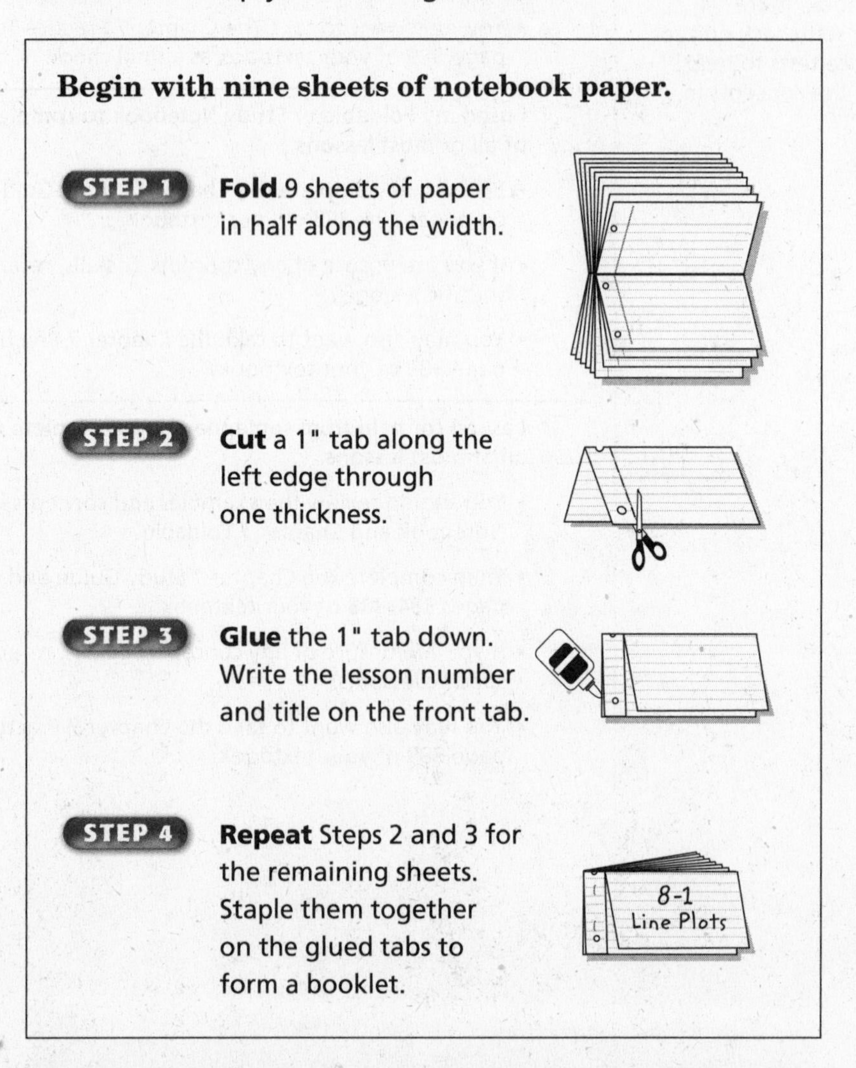

Begin with nine sheets of notebook paper.

STEP 1 **Fold** 9 sheets of paper in half along the width.

STEP 2 **Cut** a 1" tab along the left edge through one thickness.

STEP 3 **Glue** the 1" tab down. Write the lesson number and title on the front tab.

STEP 4 **Repeat** Steps 2 and 3 for the remaining sheets. Staple them together on the glued tabs to form a booklet.

NOTE-TAKING TIP: When you take notes, it is sometimes helpful to make a graph, diagram, picture, chart, or concept map that presents the information introduced in the lesson.

BUILD YOUR VOCABULARY

This is an alphabetical list of new vocabulary terms you will learn in Chapter 8. As you complete the study notes for the chapter, you will see Build Your Vocabulary reminders to complete each term's definition or description on these pages. Remember to add the textbook page number in the second column for reference when you study.

Vocabulary Term	Found on Page	Definition	Description or Example
analyze			
bar graph			
biased sample			
cluster			
data			
histogram			
inferences			
leaf			
line graph			
line plot			

(continued on the next page)

Vocabulary Term	Found on Page	Definition	Description or Example
mean			
measures of central tendency			
median			
mode			
outlier			
population			
random sample			
range			
scatter plot			
statistics			
stem			
stem-and-leaf plot			
survey			
unbiased sample			

8-1 Line Plots

MAIN IDEA

- Display and analyze data using a line plot.

BUILD YOUR VOCABULARY (pages 173–174)

Statistics deals with collecting, organizing, and interpreting **data.**

A **line plot** is a diagram that shows the data on a number line.

Data that is grouped closely together is called a **cluster.**

Outliers are numbers that are quite separated from the rest of the data in a data set.

EXAMPLE Display Data Using a Line Plot

1 **PRESIDENTS The table below shows the ages of the U.S. presidents at the time of their inaugurations. Make a line plot of the data.**

Age at Inauguration														
57	51	54	56	61	61	49	49	55	52	57	64	50	51	69
57	50	47	54	64	58	48	55	51	46	57	65	55	60	54
61	52	54	62	68	54	56	42	43	46	51	55	56		

Step 1 Draw a number line. Use a scale of 40 to 70 and an interval of 5.

Step 2 Place an × above the number that represents the age of each U.S. president.

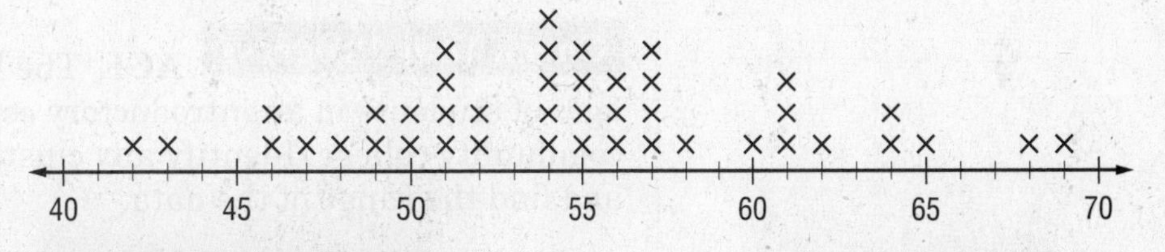

Check Your Progress

STUDY TIME The table at the right shows the number of minutes each student in a math class spent studying the night before the last math exam. Make a line plot of the data.

Minutes Studying			
36	42	60	35
70	48	55	32
60	58	42	55
38	45	60	50

FOLDABLES

ORGANIZE IT

Write a set of data that could be displayed in a line plot. Under the lab for Lesson 8-1, display the data in a line plot.

BUILD YOUR VOCABULARY (pages 173–174)

The **range** is the **difference** between the greatest and least numbers in the data set. When you **analyze** data, you use observations to describe and compare data.

EXAMPLE Use a Plot to Analyze Data

REMEMBER IT

A line plot does not need to start at 0, but you cannot leave out numbers on the number line when there are no x's above them.

2 CLIMATE **The line plot shows the number of inches of precipitation that fell in several cities west of the Mississippi River during a recent year. Identify any clusters, gaps, and outliers, and find the range of the data.**

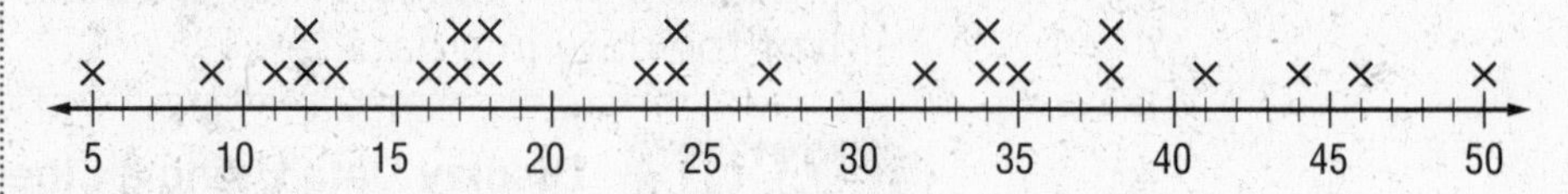

There are data clusters between ______ and 13 inches and

between 16 and ______ inches. There are gaps:

between 18 and ______; between ______ and 32.

Since ______ and 50 are apart from the rest of the data, they could be outliers.

The range is ______ – ______ or ______ inches.

Check Your Progress **AGE** The line plot below shows the ages of students in an introductory computer course at the local community college. Identify any clusters, gaps, and outliers, and find the range of the data.

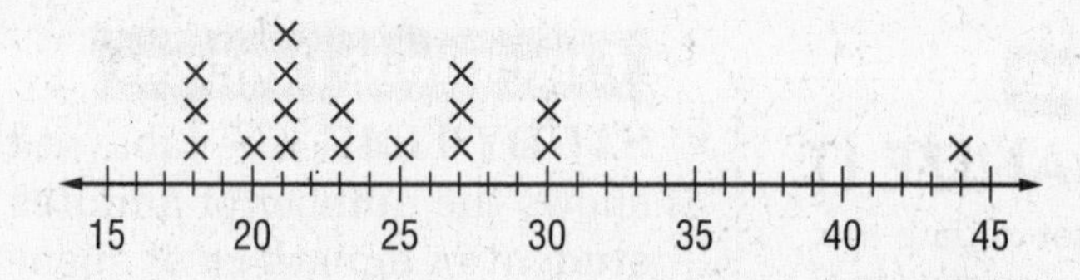

HOMEWORK ASSIGNMENT

Page(s):

Exercises:

8–2 Measures of Central Tendency and Range

MAIN IDEA

- Describe a set of data using mean, median, mode, and range.

BUILD YOUR VOCABULARY (pages 173–174)

Measures of central tendency can be used to describe the **center** of the data.

The **mean** of a set of data is the sum of the data divided by the number of items in the data set.

EXAMPLE Find the Mean

1 ANIMALS **The table below shows the number of species of animals found at 30 major zoos across the United States. Find the mean.**

Number of Species in Major U.S. Zoos				
300	400	283	400	175
617	700	700	715	280
800	290	350	133	400
195	347	488	435	640
232	350	300	300	400
705	400	800	300	659

Source: *The World Almanac*

$$\text{mean} = \frac{300 + 400 + \square + \ldots + \square}{30}$$

← sum of data

← number of data items

The mean number of species of animals is ______.

Check Your Progress **SLEEP** The table below shows the results of a survey of 15 middle school students concerning the number of hours of sleep they typically get each night. Find the mean.

Nightly Hours of Sleep				
7	8	6	7	8
9	5	6	7	7
8	6	7	8	8

FOLDABLES

ORGANIZE IT

Under the tab for Lesson 8-2, define and differentiate between mean, median, and mode.

BUILD YOUR VOCABULARY (pages 173–174)

The **median** of a set of data is the middle number of the ordered data, or the mean of the middle two numbers.

The **mode** or modes of a set of data is the number or numbers that occur most often.

EXAMPLE Find the Mean, Median, and Mode

2 OLYMPICS The table below shows the number of gold medals won by each country participating in the 2002 Winter Olympic games. Find the mean, median, and mode of the data.

2002 Winter Olympics: Gold Medals Won				
12	6	4	3	0
10	6	4	2	3
11	2	3	4	2
1	1	0	2	2
1	0	0	0	0

mean: sum of data divided by ______, or ______

median: 13th number of the ______ data, or ______

mode: number appearing ______ often, or ______

Check Your Progress **PETS** The table below shows the number of pets students in an art class at Green Hills Middle School have at home. Find the mean, median, and mode of the data.

Pets			
0	2	1	0
1	3	5	2
0	1	0	2
3	1	2	0

EXAMPLE

3 TEST EXAMPLE The average weight in pounds of several breeds of dogs is listed below.

15, 45, 26, 55, 15, 30

If the average weight of the Golden Retriever, 70 pounds, is added to this list, which of the following statements would be true?

A The mode would increase.

B The median would decrease.

C The median would increase.

D The mean would decrease.

Read the Item

You are asked to identify which statement would be true if the data value ______ was added to the data set.

Solve the Item

Use number sense to eliminate possibilities.

The mode, ______, will remain unchanged since the new data value occurs only once. So, eliminate choice ______.

Since the new data value is ______ than each value in the data set, neither the mean nor median will decrease. So, eliminate choices B and ______.

Since 70 is greater than each value in the data set, the median will now ______. So, the answer is ______.

Check Your Progress If the average weight of the Chihuahua, 4 pounds, is added to the list above, which of the following statements would be true?

F The mean would decrease.

G The mode would decrease.

H The median would stay the same.

J The mean would increase.

HOMEWORK ASSIGNMENT

Page(s):

Exercises:

8–3 Stem-and-Leaf Plots

MAIN IDEA

- Display and analyze data in a stem-and-leaf plot.

BUILD YOUR VOCABULARY (pages 173–174)

In a **stem-and-leaf plot**, the data are organized from ________ to ________.

The digits of the ________ place value usually form the **leaves** and the next place-value digits form the **stems**.

FOLDABLES

ORGANIZE IT

Under the tab for Lesson 8-3, give an example of a set of data for which a stem-and-leaf plot would be appropriate. Draw the stem-and-leaf plot.

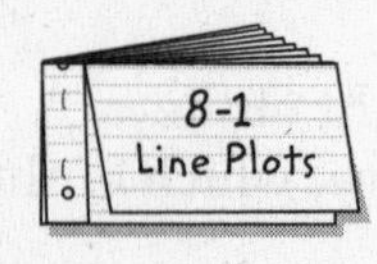

EXAMPLE Display Data in a Stem-and-Leaf Plot

1 BASEBALL **The table below shows the number of home runs that Babe Ruth hit during his career from 1914 to 1935. Make a stem-and-leaf plot of the data.**

Home Runs			
0	54	25	46
4	59	47	41
3	35	60	34
2	41	54	6
11	22	46	
29	46	49	

Step 1 The digits in the ________ place value will form the leaves and the remaining digits will form the ________. In these data, ____ is the least value, and ____ is the greatest. So, the ones digit will form the ________ and the ________ digit will form the stems.

Step 2 List the stems 0 to ______ in order from least to greatest in the *Stem* column. Write the leaves, the ______ digits of the home runs, to the ______ of the corresponding stems.

Step 3 Order the leaves and write a *key* that explains how to read the stems and leaves

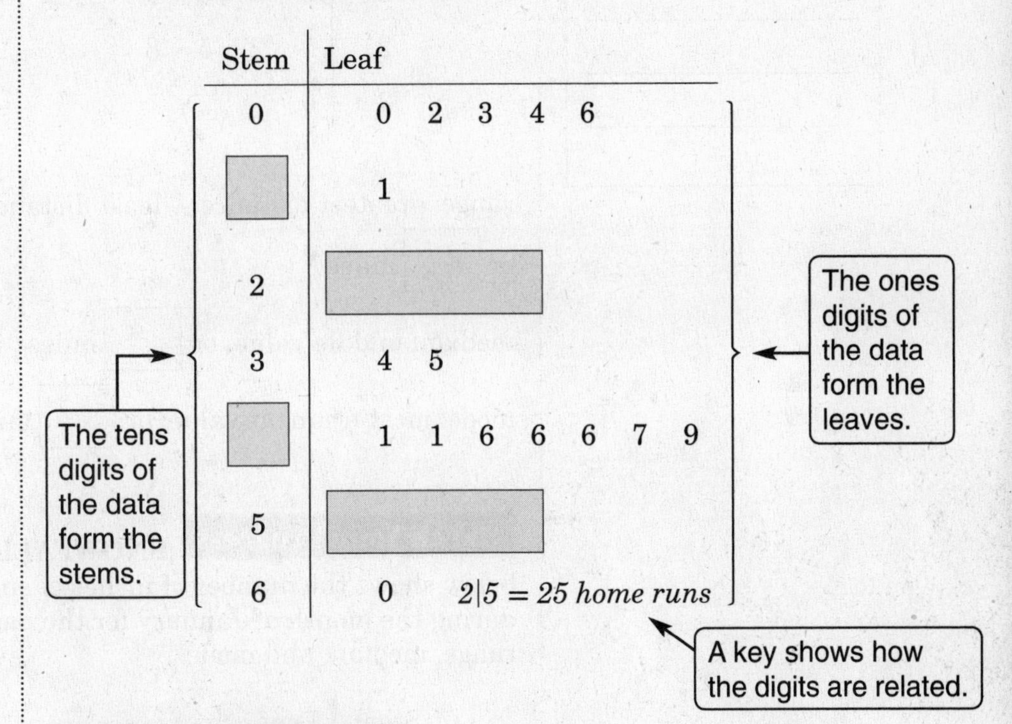

Check Your Progress **BUSINESS** The table shows the number of hours several business men and women spent aboard an airplane. Make a stem-and-leaf plot of the data.

Hours Aboard an Airplane						
4	18	0	23	12	7	9
35	14	6	11	21	19	6
15	26	9	0	13	22	10

EXAMPLE Describe Data

WRITE IT

Explain how to find how many items are on a stem-and-leaf plot.

______

______

______

______

2 FITNESS The stem-and-leaf plot below shows the number of miles that Megan biked each day during July. Find the range, median, and mode of the data.

Stem	Leaf
0	5 5 5 6
1	0 0 0 0 1 2 2 5 8 8 9
2	1 2 5 8
3	0

2|5 = 25 miles

range: greatest distance – least distance = ☐ – ☐ or ☐ miles

median: middle value, or ☐ miles

mode: most frequent value, or ☐ miles

Check Your Progress **SNOWFALL** The stem-and-leaf plot below shows the number of inches of snow that fell in Hightown during the month of January for the past 15 years. Find the range, median, and mode.

Stem	Leaf
0	1 3 5 7 9
1	0 0 0 2 4 4 7 8
2	2 6

1|2 = 12 inches

EXAMPLE Effects of Outliers

3 ANIMALS The average life span of several animal species is shown in the stem-and-leaf plot. Which measure of central tendency is most affected by the inclusion of the outlier?

Animals' Life Spans

Stem	Leaf
0	3 4 6 8
1	0 0 2 2 2 5 5 6 8
2	0 0 0 0 2
3	
4	0

1|0 = 10 years

The mode, ______, is not affected by the inclusion of the outlier, ______.

Calculate the mean and median each without the ______, 40. Then calculate them including the outlier and compare.

without the outlier

mean: $\frac{3 + 4 + \ldots + 22}{18} \approx 12.4$

median: ______

including the outlier

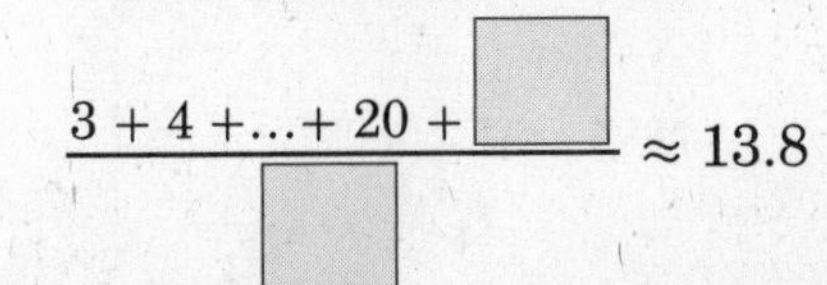

median: ______

The mean increased by 13.8 – 12.4, or ______, while the median increased by 15 – 13.5, or ______. So, the ______ is most affected by the inclusion of the outlier.

Check Your Progress

TEST SCORES The test scores earned by a class of middle school math students on a chapter test are shown. Which measure of central tendency is most affected by the inclusion of the outlier?

Test Scores

Stem	Leaf
5	8
6	
7	5 6 7 9
8	0 0 1 2 2 5 5 6 6 7
9	0 2 3 3 3 4 4 6

7|5 = 75 points

HOMEWORK ASSIGNMENT

Page(s):

Exercises:

8-4 Bar Graphs and Histograms

MAIN IDEA

- Display and analyze data using bar graphs and histograms.

BUILD YOUR VOCABULARY (pages 173–174)

A **bar graph** is one method of ________ data by using solid bars to represent quantities.

EXAMPLE Display Data Using a Bar Graph

1 TOURISM **Make a bar graph to display the data in the table below.**

Country	Vacation Days per Year
Italy	42
France	37
Germany	35
Brazil	34
United Kingdom	28
Canada	26
Korea	25
Japan	25
United States	13

Source: *The World Almanac*

FOLDABLES

ORGANIZE IT

Under the tab for Lesson 8-4, draw a sketch of a bar graph and a histogram and describe their similarities and differences.

Step 1 Draw and label the axes. Then choose a ________ on the vertical axis so that it includes all of the vacation days per year.

Step 2 Draw a ________ to represent each category.

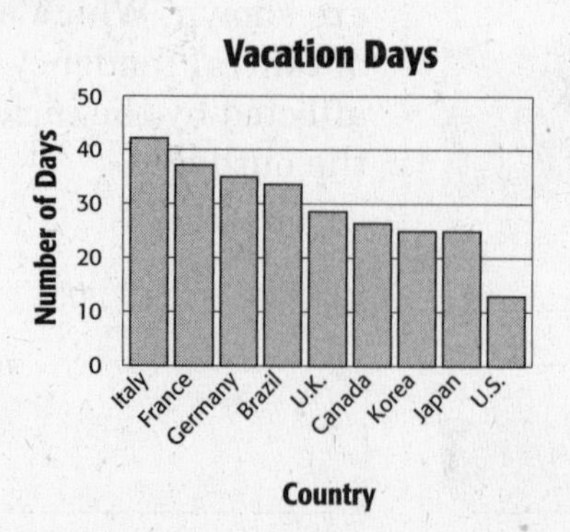

Check Your Progress

SPORTS The table shows the average number of miles run each day during training by members of the cross country track team. Make a bar graph to display the data.

Runner	Miles
Bob	9
Tamika	12
David	14
Anne	8
Jonas	5
Hana	10

BUILD YOUR VOCABULARY (pages 173–174)

A **histogram** is a special kind of ______ graph that uses bars to represent the frequency of numerical data that have been organized in ______.

WRITE IT

Explain when you would use a bar graph and when you would use a histogram.

______

______

______

______

EXAMPLE Display Data Using a Histogram

2 BASKETBALL **The number of wins for 29 teams of a basketball league for a season have been organized into a frequency table. Make a histogram of the data.**

Number of Wins	Frequency
11–20	3
21–30	4
31–40	4
41–50	10
51–60	8

(continued on the next page)

Step 1 Draw and horizontal and axes. Add a .

Step 2 Draw a bar to represent the of each interval.

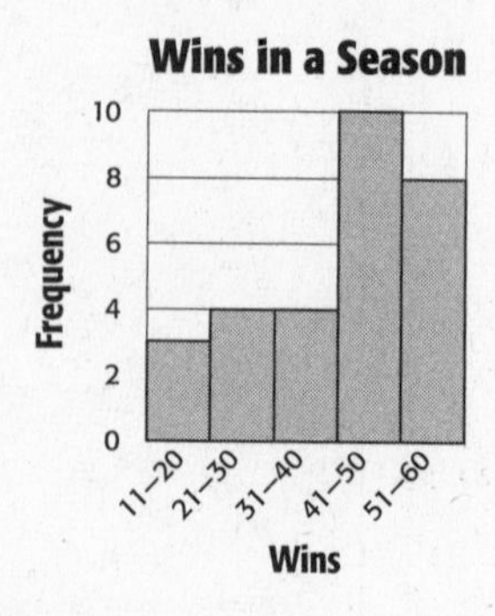

Check Your Progress **SPEED** The speeds of cars on a stretch of interstate are clocked by a police officer and have been organized into a frequency table. Make a histogram of the data.

Speed (mph)	Frequency
50–59	2
60–69	14
70–79	18
80–89	3

EXAMPLES Analyze Data to Make Inferences

DINING OUT The bar graph shows the number of times people dine out each month.

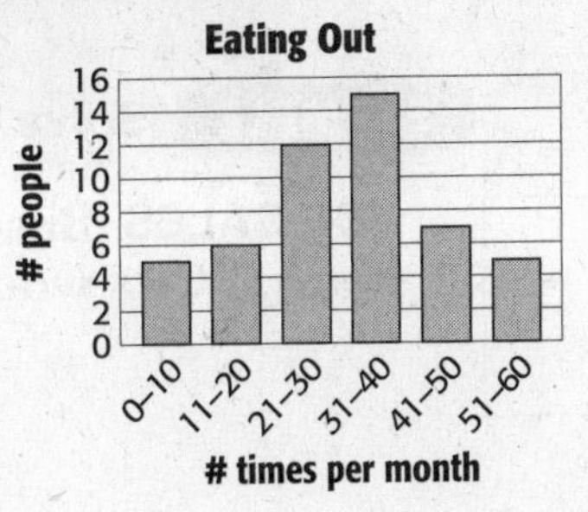

3 How many people are represented in the histogram? Justify your answer.

Find the sum of the heights of the bars in the histogram.

5 + ☐ + ☐ + 15 + 7 + ☐ = ☐

4 What percent of people surveyed ate out more than 40 times per month?

$\frac{7 + 5}{50} = \frac{\square}{50}$ ← number of people who ate out more than 40 times
← total number of people surveyed

$\frac{12}{50} = \square$ Write the fraction as a decimal.

$0.24 = \square$ Write the decimal as a percent.

So, 24% of the people surveyed ate out more than 40 times per month.

Check Your Progress

HOUSING The bar graph shows the number of houses sold in various price ranges.

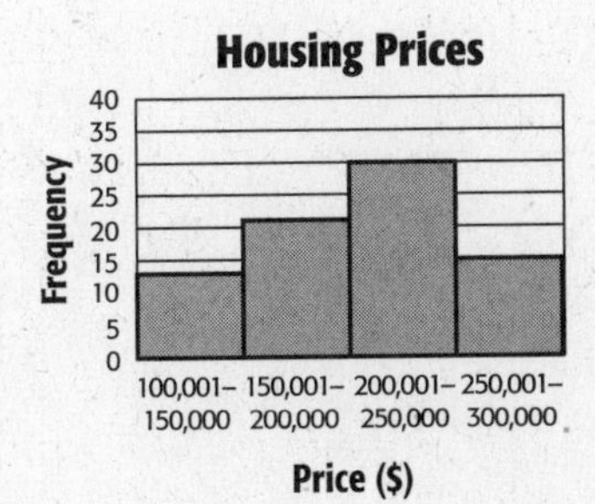

a. How many houses are represented in the histogram?

b. What percent of houses were sold for more than $200,00

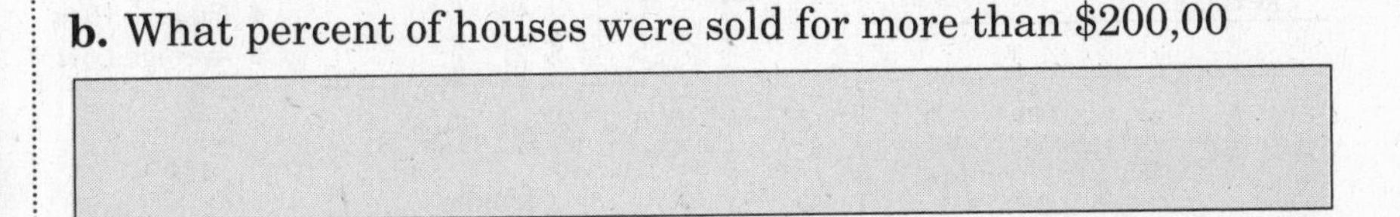

HOMEWORK ASSIGNMENT

Page(s):

Exercises:

8–5 Problem-Solving Investigation: Use a Graph

EXAMPLE Solve Problems by Using a Graph

MAIN IDEA

- Solve problems by using a graph.

VCR SALES Based on the information in the graph, how many VCRs would you expect to be sold in 2012?

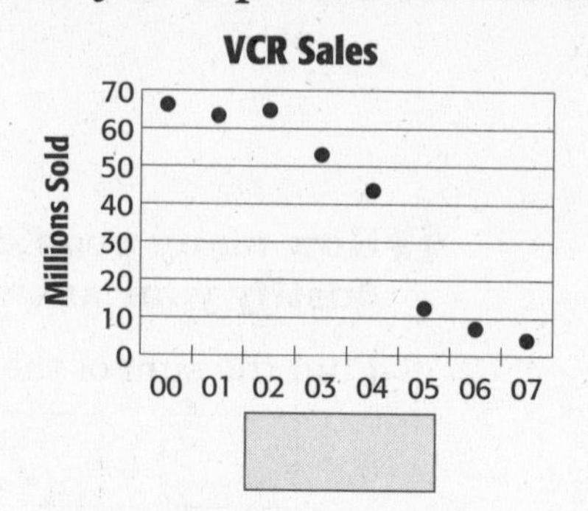

UNDERSTAND You know that the graph shows a rapid downward trend. You need to determine how many VCRs would be expected to be sold in 2012.

PLAN Look at the trend of the graph. Predict the number of VCR sales in 2012.

SOLVE If the trend continues, no VCRs will be expected to be sold in 2012.

CHECK The graph rapidly decreases. The answer is reasonable.

The graph shows a rapid ________ trend. If it continued, ________ VCRs would be sold in ________.

Check Your Progress **TEMPERATURE** Refer to the graph below. Suppose the trends continue. Predict the average high temperature for the month of August.

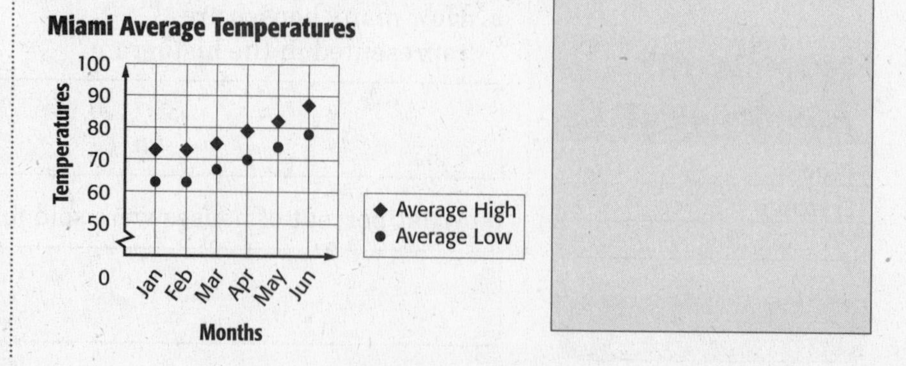

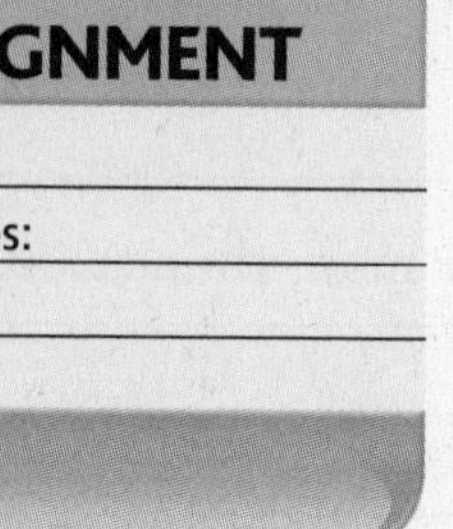

HOMEWORK ASSIGNMENT

Page(s):

Exercises:

8–6 Using Graphs to Predict

BUILD YOUR VOCABULARY (pages 173–174)

Line graphs can be useful in predicting ________ events when they show trends over ________.

MAIN IDEA

- Analyze line graphs and scatter plots to make predictions and conclusions.

EXAMPLE Use a Line Graph to Predict

1 **TYPING The line graph shows the time it has taken Enrique to type a class paper so far. The paper is 600 words long. Use the graph to predict the total time it will take him to type his paper.**

By looking at the pattern in the graph, you can predict that it will take Enrique about ________ minutes to type his 600-word paper.

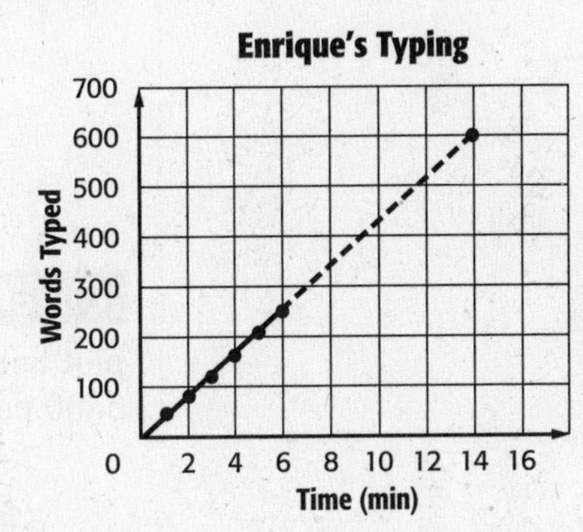

Check Your Progress **TRAVEL** During a recent road trip, Helen kept track of the number of miles traveled after each hour of travel time was completed. The table shows her information. Use the line graph to predict how far Helen will travel in 12 hours of travel time.

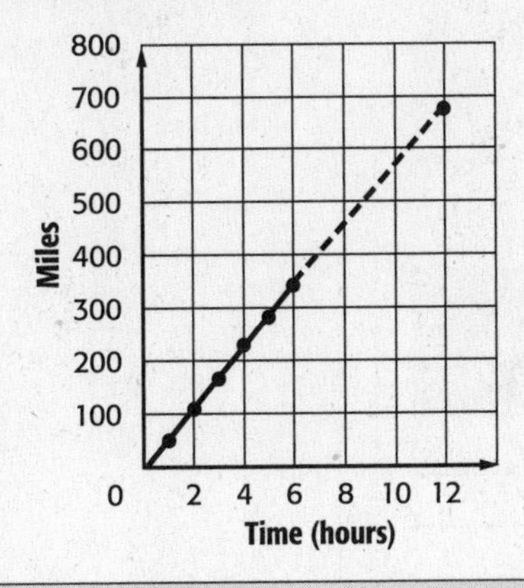

FOLDABLES

ORGANIZE IT

Under the tab for Lesson 8-6, include an example of a line graph and explain how it can be used to make predictions.

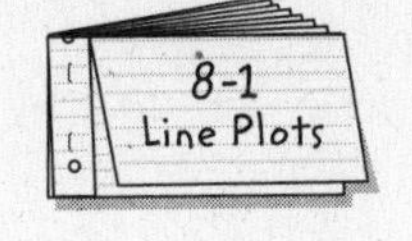

WRITE IT

Explain how a line graph can help you to make a prediction.

____

____

____

____

____

BUILD YOUR VOCABULARY (pages 173–174)

A **scatter plot** displays two sets of data on the same graph and are also useful in making ________.

EXAMPLE Use a Scatter Plot to Predict

2 POLLUTION **The scatter plot shows the number of days that a city failed to meet air quality standards from 2000 to 2008. Use it to predict the number of days of bad air quality in 2014.**

By looking at the pattern, you can predict that the number of days of bad air quality in 2014 will be about ________ days.

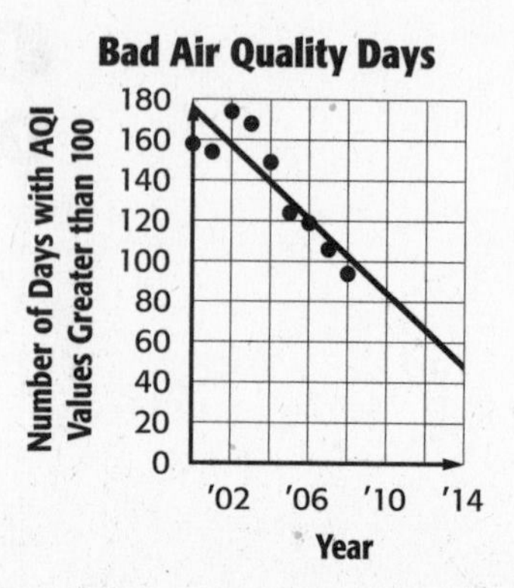

Check Your Progress **GAS MILEAGE** Use the scatter plot below to predict the gas mileage for a car weighing 5500 pounds.

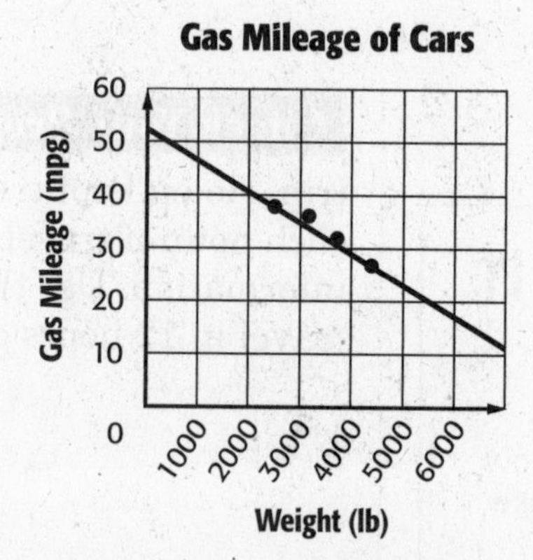

HOMEWORK ASSIGNMENT

Page(s):

Exercises:

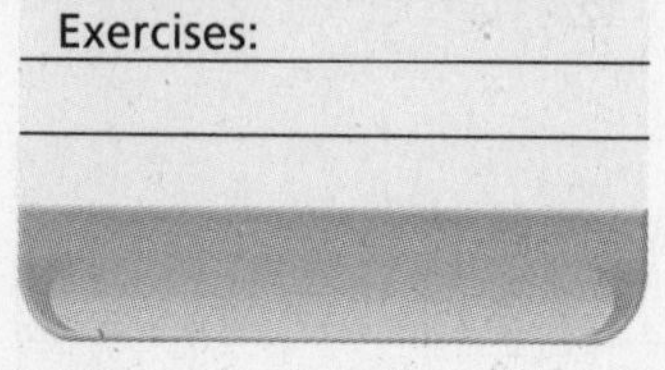

8–7 Using Data to Predict

Main Idea

- Predict actions of a larger group by using a sample.

Build Your Vocabulary (pages 173–174)

A **survey** is designed to collect ______ about a specific group of people, called the **population**.

Foldables

Organize It

Under the tab for Lesson 8-7, give examples about using statistics to predict.

8-1	8-2	8-3
8-4	8-5	8-6
8-7	8-8	8-9

EXAMPLE

1 PETS The table shows the results of a survey in which people were asked whether their house pets watch television. There are 540 students at McCloskey Middle School who own pets. Predict how many of them would say their pets watch TV.

Does your pet watch television?	
Response	**Percent**
yes	38%
no	60%
don't know	2%

You can use the percent proportion and the survey results to predict the number of people who said their pets watch TV.

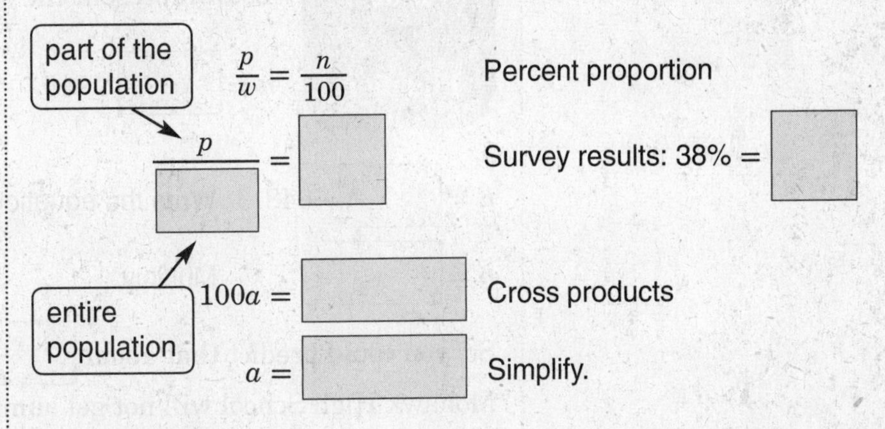

About ______ of the people surveyed said that their pets watch television.

REVIEW IT

Solve the proportion $\frac{7}{9} = \frac{x}{27}$.

Check Your Progress **VIDEO GAMES** In a survey of middle school students, 32% responded that playing video games was their favorite after-school activity. Predict how many of the 260 students surveyed said that playing video games was their favorite after-school activity.

EXAMPLE

2 **SUMMER JOBS According to one survey, 25% of high school students reported they would not get summer jobs. Predict how many of the 948 students at Mohawk High School will not get summer jobs.**

You need to predict how many of the ______ students will not get summer jobs.

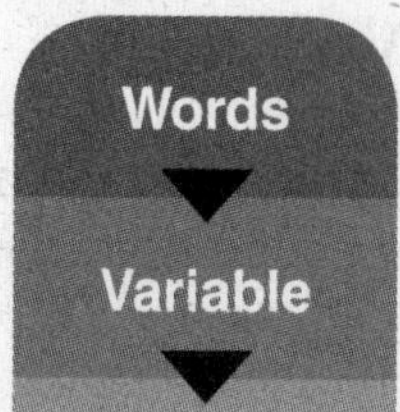

What number is 25% of 948?

Let n represent the ______.

$n =$ ______ $\cdot$ 948

$n =$ ______ $\cdot$ 948 Write the equation.

$n =$ ______ Multiply.

So, you could predict that about ______ of the students at Mohawk High School will not get summer jobs.

Check Your Progress **SEASONS** According to one survey, 31% of adults consider spring to be their favorite season of the year. Predict how many of the 525 employees of a large corporation would respond that spring is their favorite season of the year.

HOMEWORK ASSIGNMENT

Page(s):

Exercises:

8–8 Using Sampling to Predict

MAIN IDEA

- Predict the actions of a larger group by using a sample.

BUILD YOUR VOCABULARY (pages 173–174)

A **sample** is representative of a larger population. An **unbiased sample** is representative of the entire population. A **simple random sample** is the most common type of unbiased sample.

A **biased sample** occurs when one or more parts of the population are favored over others. A **convenience sample** includes members of a population who are easily accessed. A **voluntary response sample** involves only those who want to participate in sampling.

EXAMPLE Determine Validity of Conclusions

Determine whether the conclusion is valid. Justify your answer.

1 **A newspaper asks its readers to answer a poll about whether or not an issue should be on the ballot in an upcoming election. 85% of the readers who responded said that they wanted the issue on the ballot, so the newspaper printed an article saying that 85% of people want the issue on the ballot.**

The conclusion is ______. The population is restricted to readers and it is a voluntary response sample and is ______. The results of a voluntary response sample do not necessarily represent the entire ______.

Check Your Progress **Determine whether the conlusion is valid. Justify your answer.**

A coffee shop asks every tenth customer that comes in the door to identify their favorite coffee drink. 45% of the customers surveyed said the mocha coffee is their favorite drink. The manager of the store concluded that about half of the store's customers like the mocha coffee.

EXAMPLE

2 **VENDING MACHINES An office building manager interviewed 60 of their employees to determine whether or not a vending machine should be placed in the break room. 45 of the employees said yes and 15 said no. If there are 255 employees in the building, predict how many employees would like a vending machine in the break room.**

The sample is an unbiased ______ sample since employees were randomly selected. Thus, the sample is valid.

$\frac{45}{60}$ or ______% of the employees would like a vending machine in the break room. So, find 75% of ______.

$0.75 \times 255 =$ ______ 75% of 255 = 0.75 ______ 255

So, about ______ employees would like a vending machine in the break room.

Check Your Progress **CLUBS** A Spanish teacher is trying to determine if students would be interested in joining a Spanish club. She randomly asked 30 of her students. 18 of the students said yes and 12 said no. If the teacher has 105 students in her Spanish classes, predict how many would like to join a Spanish club.

HOMEWORK ASSIGNMENT

Page(s):

Exercises:

8–9 Misleading Statistics

MAIN IDEA

- Recognize when statistics and graphs are misleading.

EXAMPLE Changing the Interval of Graphs

1 BUSINESS **The line graphs below show the last 10 weeks of sales for the Crumby Cookie Bakery.**

Sales, Graph A

Sales ($): 0, 600, 800, 1,000, 1,200

Week: 1 2 3 4 5 6 7 8 9 10

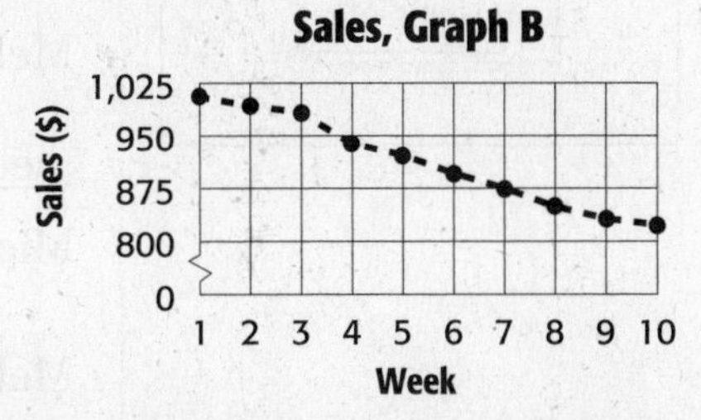

a. Do the graphs show the same data? If so, explain how the graphs differ.

The graphs show the ______ data. However, the graphs differ in that Graph ______ has greater intervals and a greater range.

b. Which graph makes it appear that the bakery's sales declined only slightly?

Graph ______ makes it appear that the sales declined only slightly even though both graphs show the same decline.

Check Your Progress SOCCER The graphs show the number of wins by four different soccer teams. Do the graphs show the same data? If so, explain how they differ.

Graph A
Wins by Soccer Teams in the Pony League

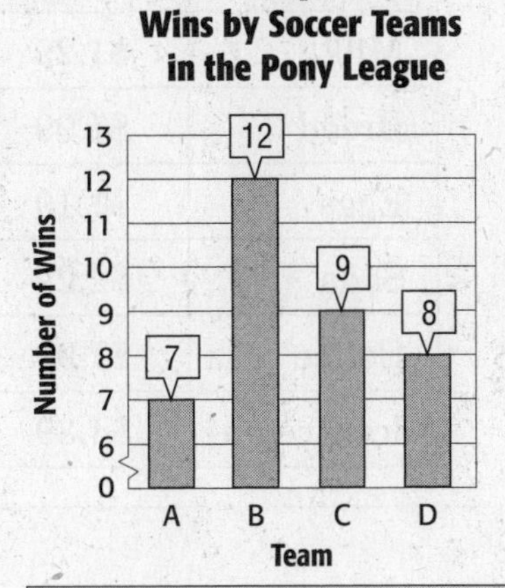

Graph B
Wins by Soccer Teams in the Pony League

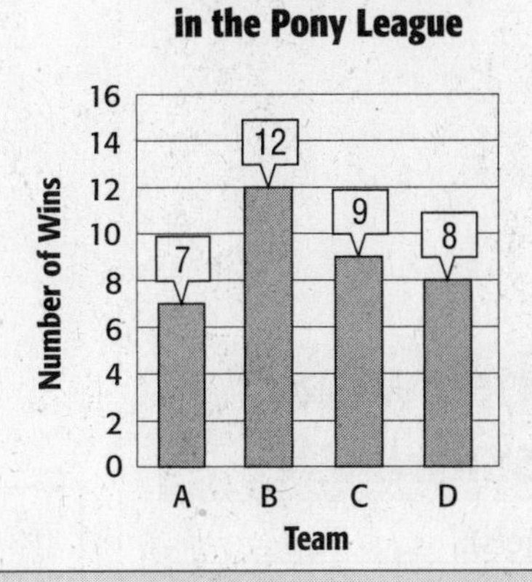

FOLDABLES

ORGANIZE IT

Under the tab for Lesson 8-9, explain how to recognize misleading graphs and statistics.

EXAMPLE Misleading Statistics

2 GRADES **Michael and Melissa both claim to be earning a C average, 70% to 79%, in their Latin class. One student is wrong. Which one? Explain how he or she is using a misleading statistic.**

mean

Michael:

Melissa:

median

Michael:

Melissa:

Test	Grade (%)	
	Michael	Melissa
1	80	88
2	76	83
3	73	75
4	70	70
5	40	60
6	25	65
7	10	62

Michael is wrong. He is using the ______ to describe his grade rather than the ______. Only Melissa's mean, or average, is 70% or better.

Check Your Progress **RETAIL SALES** Two different grocery stores each claim to have the lowest average prices. Use the table to explain their reasoning and determine which store really has the lowest average prices.

Item	Store A	Store B
Milk	$1.29	$1.34
Bread	$1.99	$1.85
Eggs	$1.19	$1.09
Soda	$2.29	$2.99
Coffee	$7.99	$5.29
Ice Cream	$4.39	$4.19

HOMEWORK ASSIGNMENT

Page(s):

Exercises:

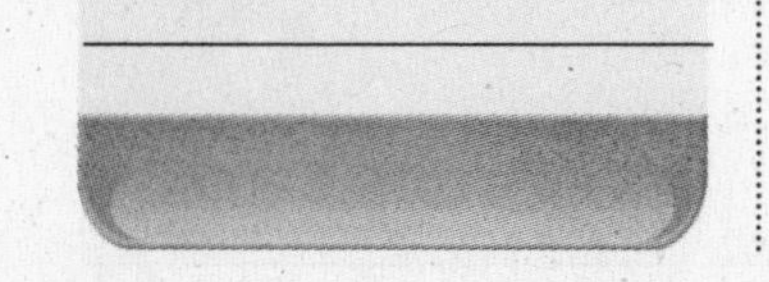

CHAPTER 8

BRINGING IT ALL TOGETHER

STUDY GUIDE

FOLDABLES	VOCABULARY PUZZLEMAKER	BUILD YOUR VOCABULARY
Use your **Chapter 8 Foldable** to help you study for your chapter test.	To make a crossword puzzle, word search, or jumble puzzle of the vocabulary words in Chapter 8, go to: glencoe.com	You can use your completed **Vocabulary Builder** *(pages 173–174)* to help you solve the puzzle.

8-1

Line Plots

The line plot shows prices for different running shoes.

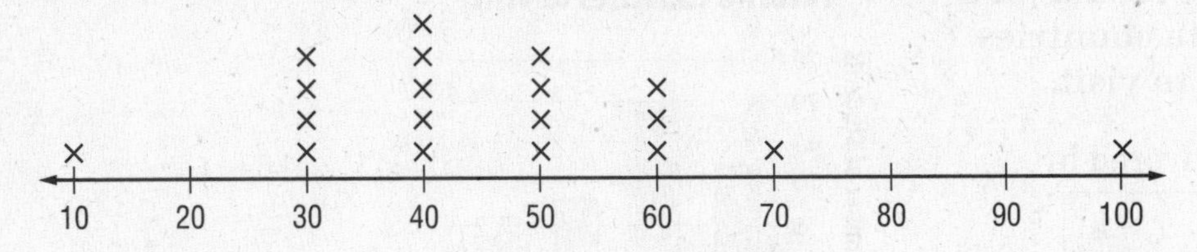

1. What is the range of the prices?

8-2

Measures of Central Tendency and Range

Find the mean, median, and mode of each set of data.

2. 2, 5, 5, 6, 8, 11, 12

3. 6, 5, 12, 34, 20, 17

8-3

Stem-and-Leaf Plots

4. The stem-and-leaf plot shows test scores for 13 students. Find the range, median, and mode of the data.

Stem	Leaf
0	7 8
1	5 5 6 9
2	0 2 2 3 3 3 4
	$1|5 = 15$

8-4 Bar Graphs and Histograms

Write *true* or *false* for each statement. If the statement is *false*, replace the underlined words with words that will make the statement true.

5. A bar graph is used to compare data.

6. A histogram shows categories on one of the axes.

8-5 Problem-Solving Investigation: Use a Graph

The graph shows the results of a survey about favorite countries students would like to visit.

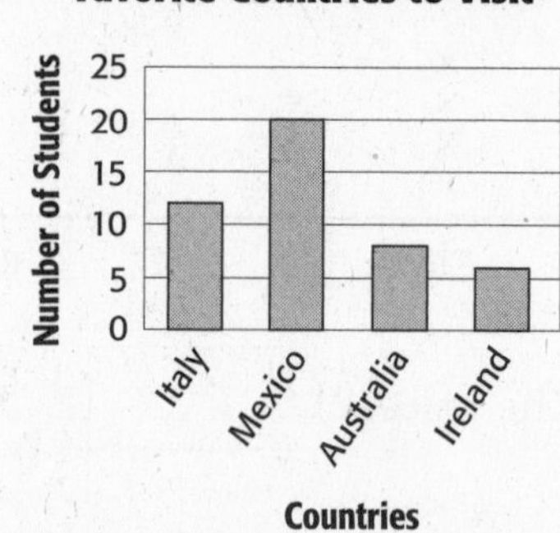

7. Which place was favored by most students?

8. Compare the number of students that would like to visit Italy versus Ireland.

8-6 Using Graphs To Predict

Refer to the graph shown.

9. Mark the City Zoo graph to show how to predict the attendance in 2005.

10. If the trend continues, predict the attendance in 2005.

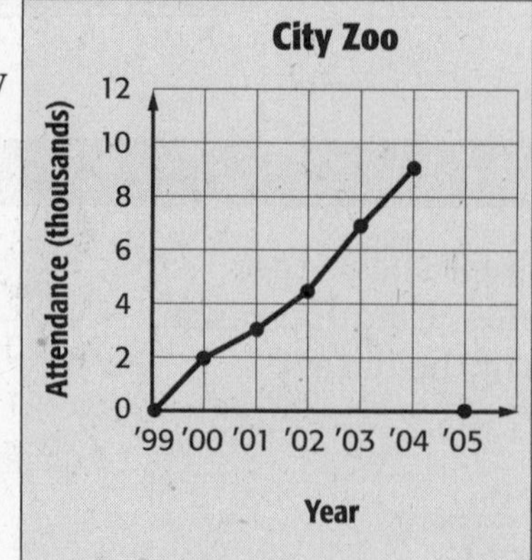

8-7 Using Data To Predict

11. **LUNCHES** A survey of 7th graders showed that 44% bring their lunch to school. Predict how many of the 450 7th graders bring their lunch to school.

12. **ZOO** A survey of zoo visitors showed that 28% chose the lion exhibit as their favorite. If 338 people visited today, predict how many would choose the lion exhibit as their favorite.

8-8 Using Sampling To Predict

Determine whether each conclusion is valid. Justify your answer.

13. A researcher randomly surveys ten employees from each department of a large company to determine the number of employees that buy their lunch in the cafeteria. Of these, 82% said they do buy their lunch in the cafeteria. The researcher concludes that most of the employees do buy their lunch in the cafeteria.

14. Every tenth customer who purchases books from an online store is asked to take a survey. The majority of those who replied said they would like more shipping options. As a result, the store adds more shipping options for their customers.

8-9 Misleading Statistics

The table lists the number of wrong answers a student had on her homework papers this year.

Wrong Answers				
1	8	2	7	2
6	8	7	2	4
7	2	5	8	6

15. Which measure of central tendency might she use to emphasize her good work?

16. Which measure of central tendency best represents her work? Explain.

ARE YOU READY FOR THE CHAPTER TEST?

Math Online

Visit **glencoe.com** to access your textbook, more examples, self-check quizzes, and practice tests to help you study the concepts in Chapter 8.

Check the one that applies. Suggestions to help you study are given with each item.

☐ **I completed the review of all or most lessons without using my notes or asking for help.**

- You are probably ready for the Chapter Test.
- You may want to take the Chapter 8 Practice Test on page 455 of your textbook as a final check.

☐ **I used my Foldables or Study Notebook to complete the review of all or most lessons.**

- You should complete the Chapter 8 Study Guide and Review on pages 450–454 of your textbook.
- If you are unsure of any concepts or skills, refer back to the specific lesson(s).
- You may want to take the Chapter 8 Practice Test on page 455 of your textbook.

☐ **I asked for help from someone else to complete the review of all or most lessons.**

- You should review the examples and concepts in your Study Notebook and Chapter 8 Foldables.
- Then complete the Chapter 8 Study Guide and Review on pages 450–454 of your textbook.
- If you are unsure of any concepts or skills, refer back to the specific lesson(s).
- You may also want to take the Chapter 8 Practice Test on page 455 of your textbook.

Student Signature

Parent/Guardian Signature

Teacher Signature

Probability

Use the instructions below to make a Foldable to help you organize your notes as you study the chapter. You will see Foldable reminders in the margin of this Interactive Study Notebook to help you in taking notes.

Begin with five sheets of $8\frac{1}{2}$" by 11" paper.

STEP 1 **Stack** 5 sheets of paper $\frac{3}{4}$ inch apart.

STEP 2 **Roll** up bottom edges so that all tabs are the same size.

STEP 3 **Crease** and staple along fold.

STEP 4 **Write** the chapter title on the front. Label each tab with a lesson number and title. Label the last tab *Vocabulary*.

NOTE-TAKING TIP: When taking notes, writing a paragraph that describes the concepts, the computational skills and the graphics will help you to understand the math in a lesson.

BUILD YOUR VOCABULARY

This is an alphabetical list of new vocabulary terms you will learn in Chapter 9. As you complete the study notes for the chapter, you will see Build Your Vocabulary reminders to complete each term's definition or description of these pages. Remember to add the textbook page number in the second column for reference when you study.

Vocabulary Term	Found on Page	Definition	Description or Example
combination			
complementary events [KAHM-pluh-MEHN-tuh-ree]			
composite events			
experimental probability [ihk-SPEHR-uh-MEHN-tuhl]			
fair game			
Fundamental Counting Principle			

Vocabulary Term	Found on Page	Definition	Description or Example
independent event			
outcome			
permutation [PUHR-myu-TAY-shuhn]			
probability [PRAH-buh-BIH-luh-tee]			
random			
sample space			
simple event			
theoretical probability [thee-uh-REHT-uh-kuhl]			
tree diagram			

9–1 Simple Events

BUILD YOUR VOCABULARY (pages 202–203)

An **outcome** is any possible ______.

A **simple event** is one ______ or a collection of outcomes.

Outcomes occur at **random** if each outcome occurs by ______.

MAIN IDEA

- Find the probability of a simple event.

KEY CONCEPT

Probability The probability of an event is a ratio that compares the number of favorable outcomes to the number of possible outcomes.

FOLDABLES On the tab for Lesson 9–1, take notes on how to find the probability of simple events. Include examples.

EXAMPLE Find Probability

1 If the spinner shown is spun once, what is the probability of its landing on an odd number?

$P(\text{odd number}) = \dfrac{\text{odd numbers possible}}{\text{total numbers possible}}$

$= \dfrac{2}{\square}$ Two numbers are odd: 1 and 3.

$=$ 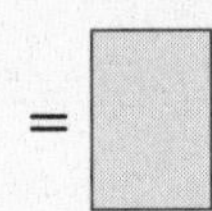 Simplify.

The probability of spinning an odd number is $\frac{1}{2}$ or ______.

Check Your Progress What is the probability of rolling a number less than three on a number cube marked with 1, 2, 3, 4, 5, and 6 on its faces?

EXAMPLE

2 GAMES A game requires spinning the spinner shown in Example 1. If the number spun is greater than 3, the player wins. What is the probability of winning the game?

Let $P(A)$ be the probability that the player will win.

$$P(A) = \frac{\text{number of favorable outcomes}}{\text{number of possible outcomes}}$$

$$= \frac{1}{4}$$

The probability of winning the game is .

REVIEW IT

Explain how to subtract a fraction from 1.

______

______

______

______

______

BUILD YOUR VOCABULARY (pages 202–203)

The sum of the probabilities of **complementary events** is 1 or 100%.

EXAMPLE

3 GAMES What is the probability of *not* winning the game described in Example 2?

$P(A) + P(\text{not } A) = 1$	Definition of complementary events
$\frac{1}{4} + P(\text{not } A) = 1$	Replace $P(A)$ with $\frac{1}{4}$.
$-\frac{1}{4} \qquad\qquad -\frac{1}{4}$	Subtract $\frac{1}{4}$ from each side.
$P(\text{not } A) =$ ______	

The probability of *not* winning the game is $\frac{3}{4}$.

Check Your Progress A game requires spinning the spinner shown in Example 1. If the number spun is less than or equal to 2, the player wins.

a. What is the probability of winning the game?

b. What is the probability of *not* winning the game?

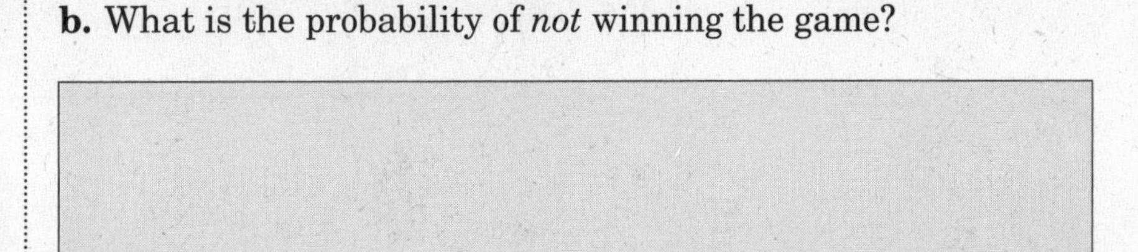

HOMEWORK ASSIGNMENT

Page(s):

Exercises:

9–2 Sample Spaces

BUILD YOUR VOCABULARY (pages 202–203)

The **sample space** is the set of all ________ outcomes.

A **tree diagram** can be used to display the ________.

MAIN IDEA

- Find sample spaces and probabilities.

EXAMPLE Find the Sample Space

1 CHILDREN **A couple would like to have two children. Find the sample space of the children's genders if having a boy is equally likely as having a girl.**

Make a table that shows all of the possible outcomes.

girl	
girl	boy
boy	
boy	girl

FOLDABLES

ORGANIZE IT

On the tab for Lesson 9–2, record what you learn about sample spaces. Explain how to find probability using a tree diagram.

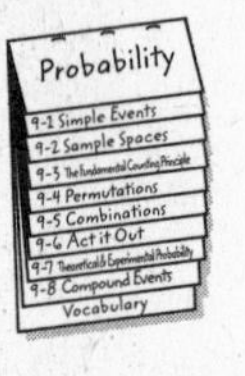

Check Your Progress **CARS** A dealer sells a car in red, black, or white. The car also can be 2-door or 4-door. Find the sample space for all possible cars available from this dealer.

EXAMPLE

2 TEST EXAMPLE Amy was trying to decide what kind of sandwich to make. She had two kinds of bread, wheat and sourdough. And she had three kinds of lunchmeat, ham, turkey, and roast beef. Which list shows all the possible bread-lunchmeat combinations?

A

Outcomes	
wheat	ham
sourdough	turkey
wheat	turkey
sourdough	ham

B

Outcomes	
wheat	ham
wheat	turkey
wheat	roast beef

C

Outcomes	
wheat	ham
wheat	turkey
wheat	roast beef
sourdough	ham
sourdough	turkey
sourdough	roast beef

D

Outcomes	
wheat	turkey
sourdough	turkey
wheat	turkey
sourdough	ham
wheat	ham
sourdough	ham

Read the Item

There are two bread choices and three lunchmeat choices. Find all of the bread-lunchmeat combinations.

(continued on the next page)

WRITE IT

In a probability game using two counters A and B, what would the outcome BA mean?

____

____

____

____

Solve the Item

Make a tree diagram to show the sample space.

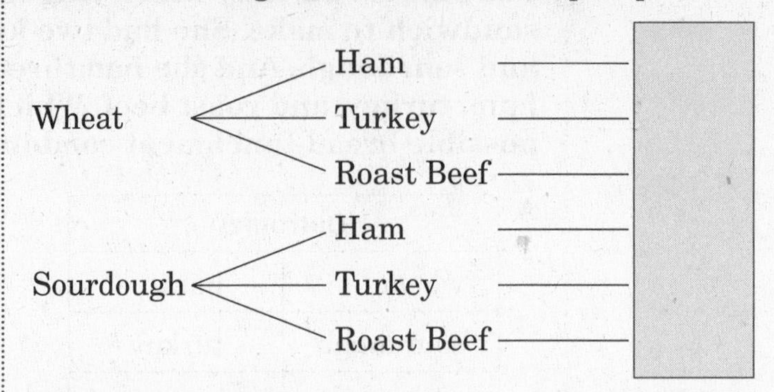

There are 6 different bread-lunchmeat combinations.

The answer is ____.

Check Your Progress **MULTIPLE CHOICE** A new car can be ordered with exterior color choices of black, red, and white, and interior color choices of tan, gray, and blue. Which list shows the different cars that are possible?

F

Outcomes	
black	tan
red	tan
white	tan
black	gray
red	gray
white	gray
black	blue
red	blue
white	blue

H

Outcomes	
black	tan
red	gray
white	blue
black	gray
red	blue
white	tan

G

Outcomes	
black	tan
red	gray
white	blue
black	gray

J

Outcomes	
black	tan
red	gray
white	blue

HOMEWORK ASSIGNMENT

Page(s):

Exercises:

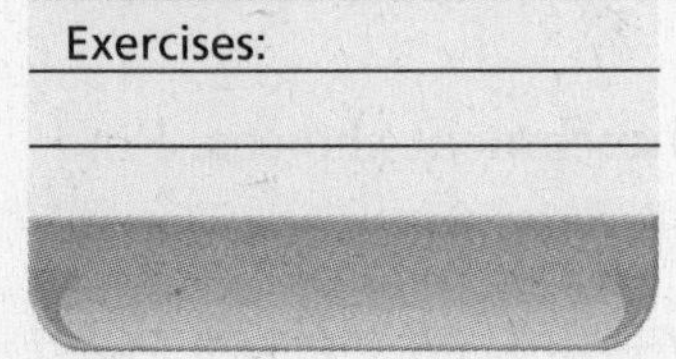

9–3 The Fundamental Counting Principle

MAIN IDEA

- Use multiplication to count outcomes and find probabilities.

BUILD YOUR VOCABULARY (pages 202–203)

You can use the **Fundamental Counting Principle** to find the number of possible outcomes in a sample space.

KEY CONCEPT

The Fundamental Counting Principle If event *M* can occur in *m* ways and is followed by event *N* that can occur in *n* ways, then the event *M* followed by *N* can occur in $m \times n$ ways.

FOLDABLES Include this concept in your notes.

EXAMPLE

1 CLOTHING **The table below shows the shirts, shorts, and shoes in Gerry's wardrobe. How many possible outfits—one shirt, one pair of shorts, and one pair of shoes—can he choose?**

Shirts	Shorts	Shoes
red	beige	black
blue	green	brown
white	blue	
yellow		

shirts × shorts × shoes = total

☐ × ☐ × ☐ = ☐

There are ☐ possible outfits that Gerry can choose.

Check Your Progress **SANDWICHES** The table below shows the types of bread, types of cheese, and types of meat that are available to make a sandwich. How many possible sandwiches can be made by selecting one type of bread, one type of cheese, and one type of meat?

Bread	Cheese	Meat
White	American	Turkey
Wheat	Swiss	Ham
Rye	Mozzarella	Roast Beef

HOMEWORK ASSIGNMENT

Page(s):

Exercises:

9-4 Permutations

MAIN IDEA

- Find the number of permutations of a set of objects and find probabilities.

BUILD YOUR VOCABULARY (pages 202–203)

A **permutation** is an ______, or listing of objects in which ______ is important.

EXAMPLE Find a Permutation

1 **BOWLING A team of bowlers has five members, who bowl one at a time. In how many orders can they bowl?**

There are ____ choices for the first bowler.

There are ____ choices for the second bowler.

There are ____ choices for the third bowler.

There are ____ choices for the fourth bowler.

There is ____ choice that remains.

$5 \cdot 4 \cdot 3 \cdot 2 \cdot 1 =$ ____

There are ____ possible arrangements, or permutations, of the five bowlers.

KEY CONCEPT

Factorial The expression *n* factorial (*n*!) is the product of all counting numbers beginning with *n* and counting backward to 1.

Check Your Progress **TRACK AND FIELD** A relay team has four members who run one at a time. In how many orders can they run?

EXAMPLE Find a Permutation

FOLDABLES

ORGANIZE IT

On the tab for Lesson 9–4, record what you learn about permutations.

2 **RAFFLE A school fair holds a raffle with 1st, 2nd, and 3rd prizes. Seven people enter the raffle, including Marcos, Lilly, and Heather. What is the probability that Marcos will win the 1st prize, Lilly will win the 2nd prize, and Heather will win the 3rd prize?**

There are ______ choices for 1st prize.

There are ______ choices for 2nd prize.

There are ______ choices for 3rd prize.

$7 \cdot 6 \cdot 5 = 210$ ← The number of permutations of 3 prizes.

There are ______ possible arrangements, or permutations, of the 3 prizes. Since there is only one way of arranging Marcos first, Lilly second, and Heather third, the probability of this event is ______.

Check Your Progress **CLUBS** The president and vice president of the French Club will be randomly selected from a jar of 24 names. Find the probability that Sophie will be selected as president and Peter selected as vice president.

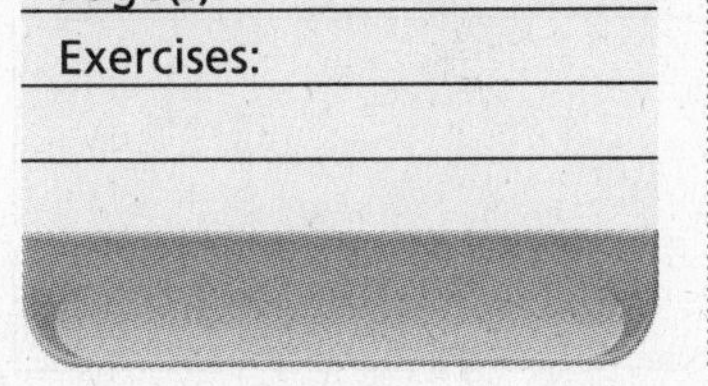

HOMEWORK ASSIGNMENT

Page(s):

Exercises:

9–5 Combinations

MAIN IDEA

- Find the number of combinations of a set of objects and find probabilities.

BUILD YOUR VOCABULARY (pages 202–203)

An arrangement, or listing, of objects in which order is ______ is called a **combination.**

FOLDABLES

ORGANIZE IT

On the tab for Lesson 9–5, record what you learn about combinations. Be sure to compare and contrast combinations and permutations.

EXAMPLE Find the Number of Combinations

1 DECORATING Ada can select from seven paint colors for her room. She wants to choose two colors to paint stripes on her walls. How many different pairs of colors can she choose?

METHOD 1 Make a list.

Number the colors 1 through 7.

1, 2	1, 5	2, 3	2, 6	3, 5	4, 5	5, 6
1, 3	1, 6	2, 4	2, 7	3, 6	4, 6	5, 7
1, 4	1, 7	2, 5	3, 4	3, 7	4, 7	6, 7

There are ______ different pairs of colors.

METHOD 2 Use a permutation.

There are $7 \cdot 6$ permutations of two colors chosen from seven. There are $2 \cdot 1$ ways to arrange the two colors.

$\frac{7 \cdot 6}{2 \cdot 1} =$ ______ $=$ ______

There are ______ different pairs of colors Ada can choose.

Check Your Progress **HOCKEY** The Brownsville Badgers hockey team has 14 members. Two members of the team are to be selected to be the team's co-captains. How many different pairs of players can be selected to be the co-captains?

REMEMBER IT

To find a combination you must divide the permutation by the number of ways you can arrange the items.

EXAMPLES

2 INTRODUCTIONS Ten managers attend a business meeting. Each person exchanges names with each other person once. How many introductions will there be?

There are $10 \cdot 9$ ways to choose 2 people.

There are $2 \cdot 1$ ways to arrange the 2 people.

$\frac{10 \cdot 9}{2 \cdot 1} = \frac{90}{2}$ or ______

There are ______ introductions.

3 If the introductions in Example 2 are made at random, what is the probability that Ms. Apple and Mr. Zimmer will be the last managers to exchange names?

Since there are ______ introductions and only one favorable outcome, the probability that Ms. Apple and Mr. Zimmer will be the last managers to exchange names is ______.

Check Your Progress

a. INTRODUCTIONS Fifteen managers attend a business meeting. Each person exchanges names with each other person once. How many introductions will there be?

b. What is the probability that Ms. Apple and Mr. Zimmer will be the last managers to exchange names?

HOMEWORK ASSIGNMENT

Page(s):

Exercises:

9–6 Problem-Solving Investigation: Act It Out

EXAMPLE Solve Using the Act It Out Strategy

MAIN IDEA

- Solve problems by acting it out.

LUNCH **Salvador is looking for his lunch money, which he put in one of the pockets of his backpack this morning. If the backpack has six pockets, what is the probability that he will find the money in the first pocket that he checks?**

UNDERSTAND You know that there are ______ pockets in Salvador's backpack and that one of the pockets contains his lunch money.

PLAN Toss a number cube several times. If the cube lands on 1, Salvador will find the money in the first pocket that he checks. If the cube lands on 2, 3, 4, 5, or 6, Salvador will not find the money in the first pocket that he checks.

SOLVE Toss the cube and make a table of the results.

Trials	1	2	3	4	5	6	7	8	9	10	11	12
Outcome	4	5	1	2	2	3	6	4	5	2	1	3

The highlighted entries show that ______ out of the 12 trials resulted in Salvador finding his lunch money in the first pocket that he checks.

So, the probability is $\frac{2}{12}$ or ______.

CHECK Repeat the experiment several times to see whether the results agree.

Check Your Progress **PHOTOGRAPHS** A photographer is taking a picture of the four members in Margaret's family. Margaret's grandmother will stand on the right. How many different ways can the photographer arrange the family members in a row for the photo?

HOMEWORK ASSIGNMENT

Page(s):

Exercises:

9–7 Theoretical and Experimental Probability

MAIN IDEA

- Find and compare experimental and theoretical probabilities.

BUILD YOUR VOCABULARY (pages 202–203)

Experimental probability is based on what ______ occurred during an experiment. **Theoretical probability** is based on what ______ happen when conducting an experiment.

FOLDABLES

ORGANIZE IT

On the tab for Lesson 9-7, take notes about theoretical and experimental probability. Be sure to describe their differences.

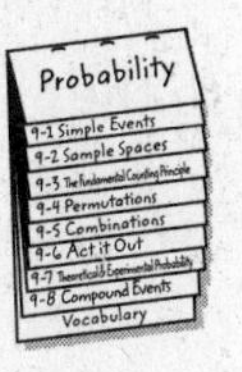

EXAMPLE Experimental Probability

1 **A spinner is spun 50 times, and it lands on the color blue 15 times. What is the experimental probability of spinning blue?**

$$P(\text{blue}) = \frac{\text{number of times } \boxed{} \text{ is spun}}{\text{number of } \boxed{} \text{ outcomes}}$$

$$= \frac{\boxed{}}{\boxed{}} \text{ or } \boxed{}$$

The experimental probability of spinning the color blue is ______.

Check Your Progress A marble is pulled from a bag of colored marbles 30 times and 18 of the pulls result in a yellow marble. What is the experimental probability of pulling a yellow marble?

EXAMPLES Experimental and Theoretical Probability

The graph shows the results of an experiment in which a number cube is rolled 30 times.

2 Find the experimental probability of rolling a 5.

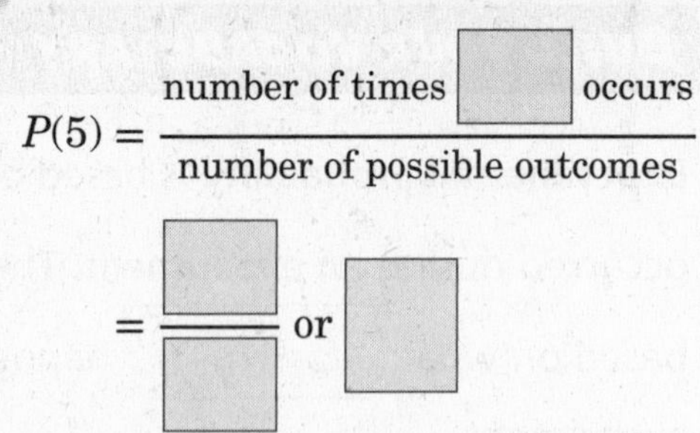

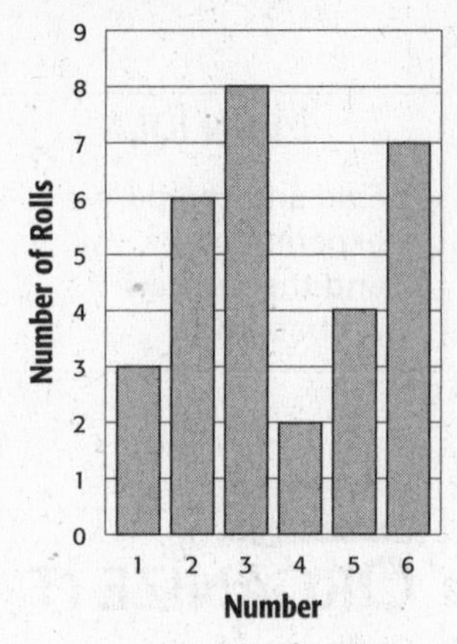

The experimental probability of rolling a ______ is ______.

3 Compare the experimental probability of rolling a 5 to its theoretical probability.

The theoretical probability of rolling a 5 on a number cube is ______. So, the theoretical probability is close to the experimental probability of ______.

Check Your Progress **The graph shows the result of an experiment in which a coin was tossed 150 times.**

a. Find the experimental probability of tossing heads for this experiment.

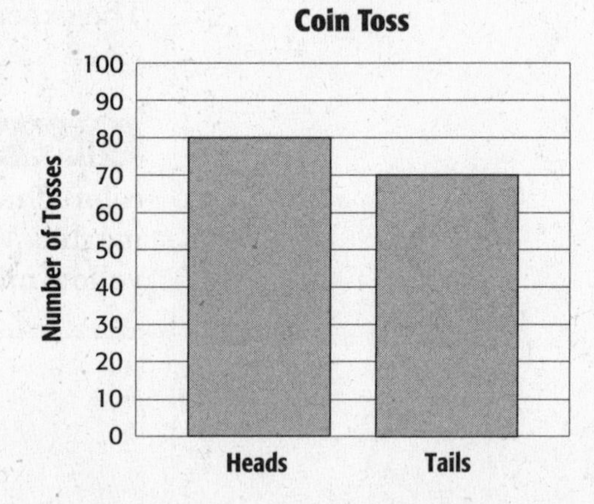

b. Compare the experimental probability of tossing heads to its theoretical probability.

HOMEWORK ASSIGNMENT

Page(s):

Exercises:

9–8 Compound Events

MAIN IDEA

- Find the probability of independent and dependent events.

BUILD YOUR VOCABULARY (pages 202–203)

A **compound event** consists of two or more events.

When choosing one event does not choosing a second event, both events are called **independent events.**

KEY CONCEPT

Probability of Two Independent Events The probability of two independent events can be found by multiplying the probability of the first event by the probability of the second event.

FOLDABLES On the tab for Lesson 9-8, give an example of finding the probability of two independent events.

EXAMPLE Independent Events

1 LUNCH **For lunch, Jessica may choose from a turkey sandwich, a tuna sandwich, a salad, or a soup. For a drink, she can choose juice, milk, or water. If she chooses a lunch and a drink at random, what is the probability that she chooses a sandwich (of either kind) and juice?**

P(sandwich) = P(juice) =

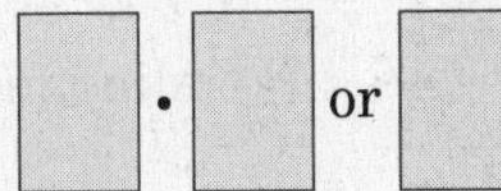

P(sandwich and juice) = · or

So, the probability that she chooses a sandwich and juice is .

Check Your Progress **SWEATS** Zachary has a blue, a red, a gray, and a white sweatshirt. He also has blue, red, and gray sweatpants. If Zachary randomly pulls a sweatshirt and a pair of sweatpants from his drawer, what is the probability that they will both be blue?

BUILD YOUR VOCABULARY (pages 202–203)

If one event affects the outcome of a second event, the events are called **dependent events.**

If two events cannot happen at the same time, then they are **disjoint events.**

EXAMPLES Dependent Events

2 SOCKS There are 4 black, 6 white, and 2 blue socks in a drawer. José randomly selects two socks without replacing the first sock. What is the probability that he selects two white socks?

$P(\text{first sock is white}) = \frac{6}{12}$ There are ______ white socks and ______ total socks.

$P(\text{second sock is white}) = \frac{5}{11}$ After one white sock is removed, there are ______ white socks and ______ total socks.

$P(\text{two white socks}) = \frac{\overset{1}{\cancel{6}}}{\underset{2}{\cancel{12}}} \cdot \frac{5}{11}$ or ______

KEY CONCEPT

Probability of two Dependent Events The probability of two dependent events, *A* and *B*, can be found by multiplying the probability of *A* by the probability of *B* after *A* occurs.

FOLDABLES On the tab for Lesson 9-8, give an example of finding the probability of two independent events.

3 Disjoint Events

MONTHS A month of the year is randomly selected. What is the probability of the month ending in the letter *Y* or the letter *R*.

They are disjoint events since it is impossible to have a month ending in both the letter *Y* and the letter *R*?

$P(\text{ending in } Y \text{ or } R) = \frac{\square}{\square}$ There are 8 months that end in *Y* or *R*. There are 12 months.

Check Your Progress

a. GAMES Janet has a card game that uses a deck of 48 cards – 16 red, 16 blue, and 16 green. If she randomly selects two cards without replacing the first, what is the probability that both are green?

b. MARBLES There are 12 yellow, 3 black, 5 red, and 8 blue marbles in a bag. Joseph randomly selects one marble from the bag. What is the probability that the marble selected will be black or red?

HOMEWORK ASSIGNMENT

Page(s):

Exercises:

CHAPTER 9

BRINGING IT ALL TOGETHER

STUDY GUIDE

FOLDABLES	VOCABULARY PUZZLEMAKER	BUILD YOUR VOCABULARY
Use your **Chapter 9 Foldable** to help you study for your chapter test.	To make a crossword puzzle, word search, or jumble puzzle of the vocabulary words in Chapter 9, go to: glencoe.com	You can use your completed **Vocabulary Builder** *(pages 202–203)* to help you solve the puzzle.

9-1 Simple Events

For Questions 1–3, a bag contains 4 green, 6 orange, and 10 purple blocks. Find each probability if you draw one block at random from the bag. Write as a fraction in simplest form.

1. P(green)

2. P(orange)

3. P(purple)

9-2 Sample Spaces

4. PHONES A phone company offers three different calling features (caller ID, call waiting, and call forward) and two different calling plans (Plan A or Plan B). Find the sample space for all possibilities of a calling feature and a calling plan.

9-3

The Fundamental Counting Principle

5. Underline the correct term to complete the sentence: The operation used in the Fundamental Counting Principle is (*addition, multiplication*).

Use the Fundamental Counting Principle to find the total number of outcomes in each situation.

6. Tossing a coin and rolling a 6-sided number cube.

7. Making a sandwich using whole wheat or sourdough bread, ham or turkey, and either cheddar, swiss, or provolone cheese.

8. Choosing a marble from a bag containing 10 differently colored marbles and spinning the spinner at the right.

9-4

Permutations

9. LETTERS How many permutations are there of the letters in the word *pizza*?

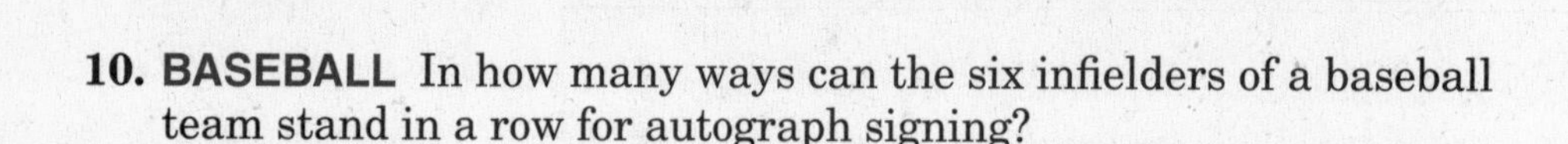

10. BASEBALL In how many ways can the six infielders of a baseball team stand in a row for autograph signing?

11. NUMBERS How many 4-digit passwords can be formed using the digits 1, 3, 4, 5, 7, and 9? Assume no number can be used more than once.

9-5

Combinations

Complete each sentence.

12. You can find the number or combinations of objects in a set by ______ the number of ______ of the entire set by the number of ways each smaller set can be arranged.

13. A ______ is an arrangement or listing in which order is not ______.

14. The burger shop offers 3 choices of condiments from the following: lettuce, onions, pickles, ketchup, and mustard. How many different combinations of condiments can you have on your burger?

9-6

Problem-Solving Investigation: Act It Out

15. TRAVEL Four friends are driving to the beach. In how many different ways can two friends sit in the front and two friends sit in the back if Raul must be the driver?

9-7

Theoretical and Experimental Probability

Underline the correct term(s) to complete each sentence.

16. The word experimental means based on (experience, theory).

17. Theoretical probability is based on what (you actually try, is expected).

18. (Experimental, theoretical) probability can be based on past performance and can be used to make predictions about future events.

Sue has 5 different kinds of shoes: sneakers, sandals, boots, moccasins, and heels.

19. If she chooses a pair each day for two weeks, and chooses moccasins 8 times, what is the experimental probability that moccasins are chosen?

20. Find the theoretical probability of choosing the moccasins.

9-8 Compound Events

State whether each sentence is *true* or *false*. If *false*, replace the underlined word to make the sentence true.

21. A compound event consists of more than one single event.

22. When the outcome of the first event does not have any effect on the second event it is called a simple event.

23. A yellow and a green cube are rolled. What is the probability that an even number is rolled on the yellow cube and a number less than 3 is rolled on the green cube?

24. There are 4 chocolate chip, 6 peanut butter, and 2 sugar cookies in a box. Malena randomly selects two cookies without replacing the first. Find the probability that she selects a peanut butter cookie and then a sugar cookie.

ARE YOU READY FOR THE CHAPTER TEST?

Math Online

Visit **glencoe.com** to access your textbook, more examples, self-check quizzes, and practice tests to help you study the concepts in Chapter 9.

Check the one that applies. Suggestions to help you study are given with each item.

☐ **I completed the review of all or most lessons without using my notes or asking for help.**

- You are probably ready for the Chapter Test.
- You may want to take the Chapter 9 Practice Test on page 503 of your textbook as a final check.

☐ **I used my Foldable or Study Notebook to complete the review of all or most lessons.**

- You should complete the Chapter 9 Study Guide and Review on pages 498–502 of your textbook.
- If you are unsure of any concepts or skills, refer back to the specific lesson(s).
- You may want to take the Chapter 9 Practice Test on page 503 of your textbook.

☐ **I asked for help from someone else to complete the review of all or most lessons.**

- You should review the examples and concepts in your Study Notebook and Chapter 9 Foldable.
- Then complete the Chapter 9 Study Guide and Review on pages 498–502 of your textbook.
- If you are unsure of any concepts or skills, refer back to the specific lesson(s).
- You may also want to take the Chapter 9 Practice Test on page 503 of your textbook.

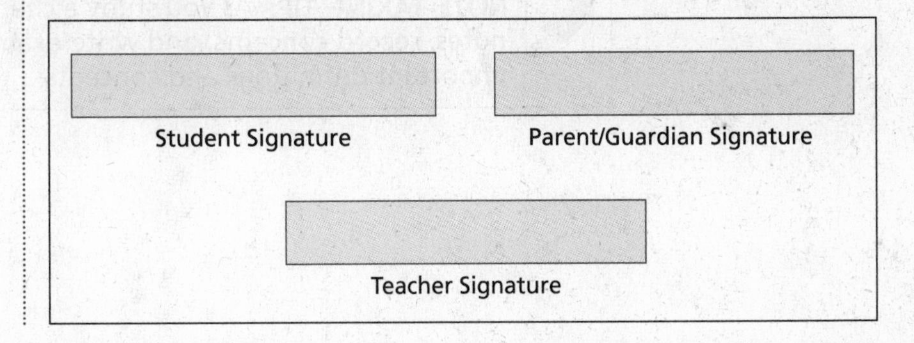

Geometry: Polygons

Use the instructions below to make a Foldable to help you organize your notes as you study the chapter. You will see Foldable reminders in the margin of this Interactive Study Notebook to help you in taking notes.

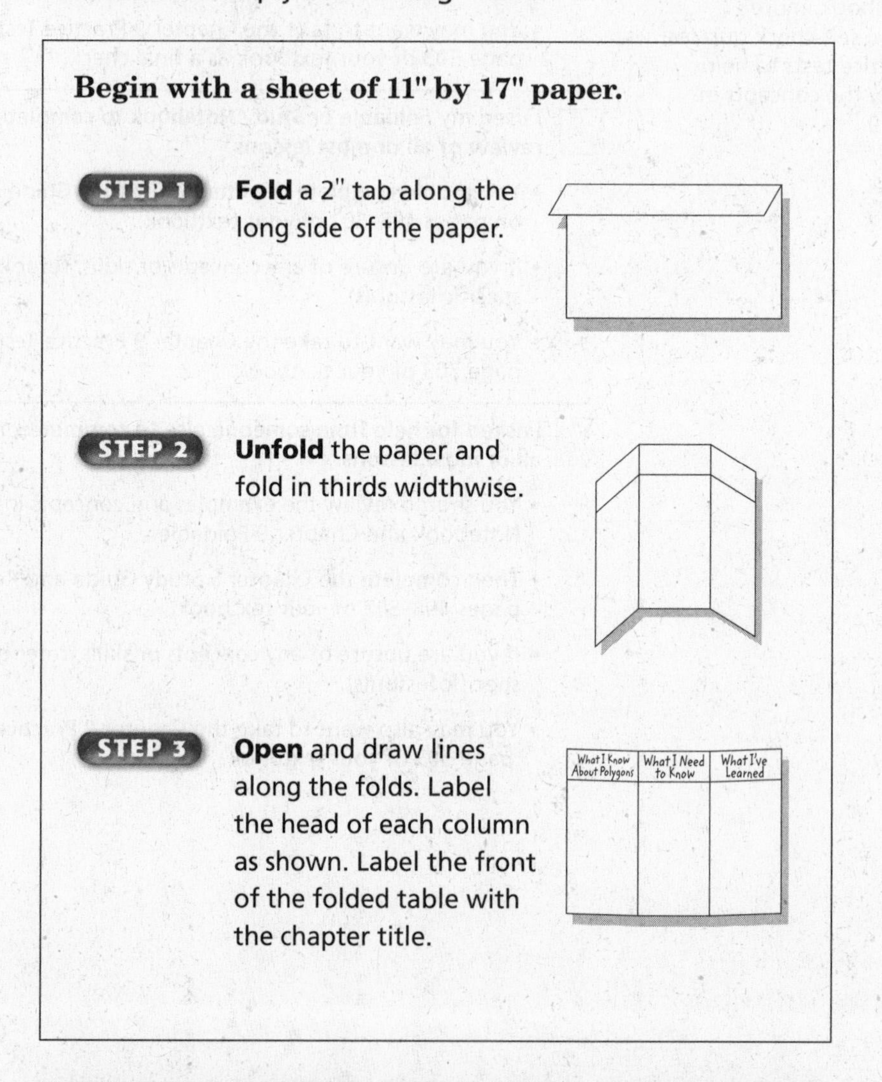

Begin with a sheet of 11" by 17" paper.

STEP 1 **Fold** a 2" tab along the long side of the paper.

STEP 2 **Unfold** the paper and fold in thirds widthwise.

STEP 3 **Open** and draw lines along the folds. Label the head of each column as shown. Label the front of the folded table with the chapter title.

NOTE-TAKING TIP: As you study a chapter, take notes, record concepts, and write examples about important definitions and concepts.

BUILD YOUR VOCABULARY

This is an alphabetical list of new vocabulary terms you will learn in Chapter 10. As you complete the study notes for the chapter, you will see Build Your Vocabulary reminders to complete each term's definition or description on these pages. Remember to add the textbook page number in the second column for reference when you study.

Vocabulary Term	Found on Page	Definition	Description or Example
acute triangle			
adjacent angles			
complementary angles			
congruent angles			
congruent segments			
equilateral [EH-kwuh-LA-tuh-rull] triangle			
indirect measurement			
isosceles [y-SAHS-LEEZ] triangle			
line symmetry			
obtuse triangle			
parallelogram			

(continued on the next page)

Vocabulary Term	Found on Page	Definition	Description or Example
quadrilateral [KWAH-druh-LA-tuh-ruhl]			
reflection			
rhombus [RAHM-buhs]			
scalene [SKAY-LEEN] triangle			
similar figures			
straight angle			
supplementary angles			
tessellation			
translation			
trapezoid [TRA-puh-ZOYD]			
vertex			
vertical angles			

10–1 Angle Relationships

MAIN IDEA

- Classify angles and identify vertical and adjacent angles.

BUILD YOUR VOCABULARY (pages 225–226)

An **angle** has two sides that share a ______ endpoint and is measured in units called **degrees**.

The ______ where the sides of an angle ______ is called the **vertex**.

EXAMPLE Naming Angles

1 Name the angle at the right.

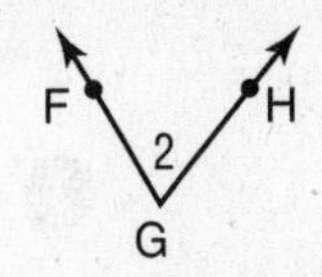

- Use the vertex as the middle letter and a point from each side.

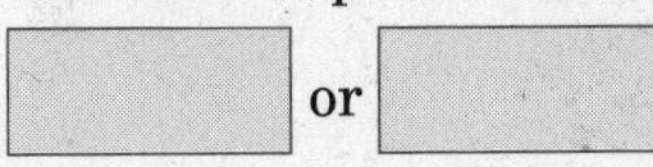

- Use the vertex only.

- Use a number.

The angle can be named in four ways:

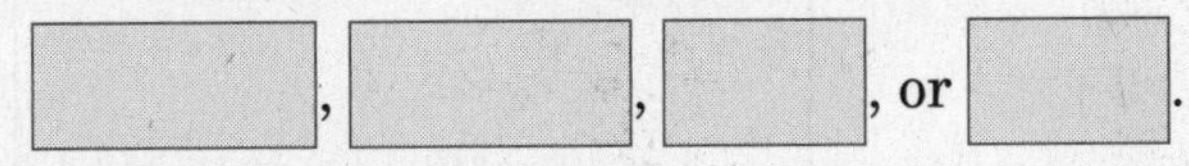

Check Your Progress Name the angle below.

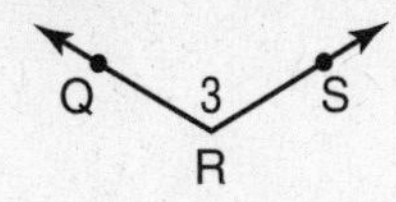

REMEMBER IT

A ray starts at a point and goes without end in one direction.

BUILD YOUR VOCABULARY (pages 225–226)

A **right angle** measures ______ 90°.

An **acute angle** measures ______ than 90°.

An **obtuse angle** measures ______ 90° and 180°.

A **straight angle** measures ______ 180°.

EXAMPLES Classify Angles

Classify each angle as *acute*, *obtuse*, *right*, or *straight*.

2

The angle is exactly ______, so it is a ______ angle.

3

The angle is ______ than 90°, so it is an ______ angle.

Check Your Progress **Classify each angle as *acute*, *obtuse*, *right*, or *straight*.**

a.

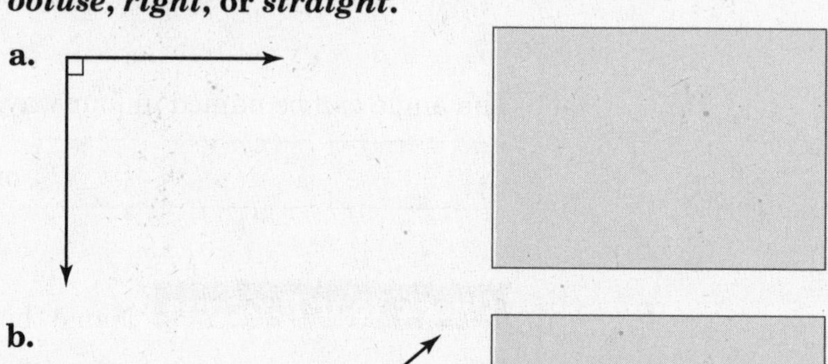

b.

BUILD YOUR VOCABULARY (pages 225–226)

Two angles that have the same ________ are **congruent**.

Two angles are **vertical** if they are ________ angles formed by the intersection of two lines.

Two angles are **adjacent** if they share a common vertex, a common ________, and do not overlap.

EXAMPLE

4 Determine if each pair of angles in the figure at the right are vertical angles, adjacent angles, or neither.

3
4
5

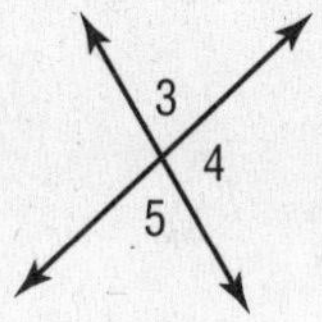

a. ∠3 and ∠5

Since ∠3 and ∠5 are opposite angles formed by the intersection of two lines, they are ________ angles.

b. ∠3 and ∠4

∠3 and ∠4 share a common vertex and side, and do not overlap. So, they are ________ angles.

c. ∠4 and ∠5

∠4 and ∠5 share a common vertex and side, and do not overlap. So, they are ________ angles.

Check Your Progress **Determine if each pair of angles in the figure at the right are vertical angles, adjacent angles, or neither.**

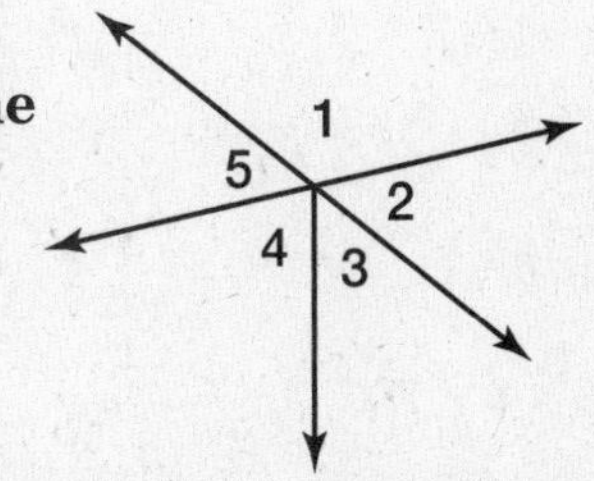

1
5
2
4
3

a. ∠1 and ∠2

____

b. ∠2 and ∠5

____

c. ∠1 and ∠4

____

10–2 Complementary and Supplementary Angles

MAIN IDEA

- Identify complementary and supplementary angles and find missing angle measures.

BUILD YOUR VOCABULARY (pages 225–226)

Complementary angles have a sum of ______.

Supplementary angles have a sum of ______.

EXAMPLES Classify Angles

Classify each pair of angles as *complementary*, *supplementary*, or *neither*.

1

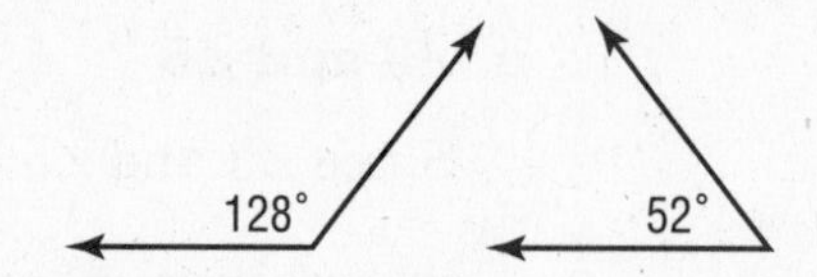

______ $+ 52° =$ ______

So, the angles are ______.

2

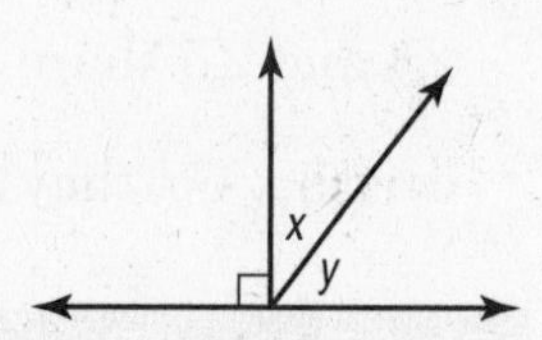

$\angle x$ and $\angle y$ form a ______ angle.

So, the angles are ______.

Check Your Progress **Classify each pair of angles as *complementary*, *supplementary*, or *neither*.**

a.

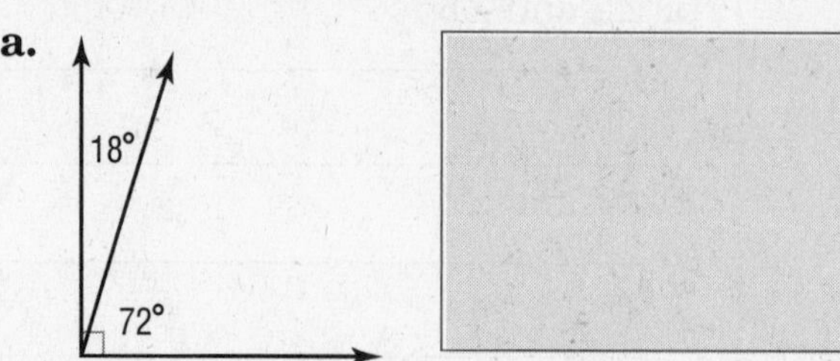

REMEMBER IT

When two angles are congruent, the measure of the angles are equal.

b.

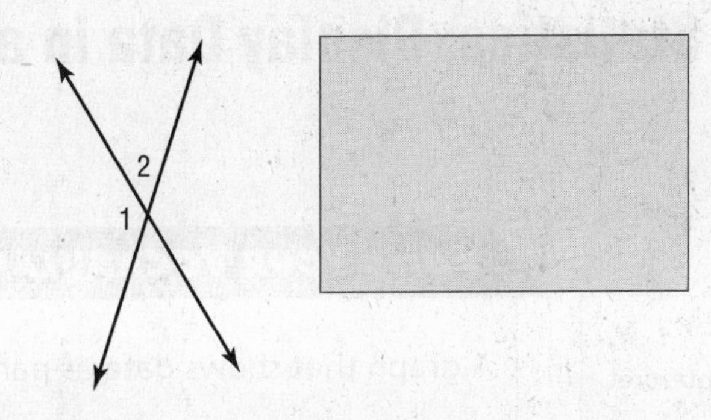

EXAMPLE Find a Missing Angle Measure

3 **Angles PQS and RQS are supplementary. If $m\angle PQS = 56°$, find $m\angle RQS$.**

Since $\angle PQS$ and $\angle RQS$ are supplementary, $m\angle PQS + m\angle RQS = 180°$.

$m\angle PQS + m\angle RQS = 180$ — Write the equation.

$\square + m\angle RQS = \square$ — Replace $m\angle PQS$ with $\square$.

$-56 \qquad -56$ — Subtract $\square$ from each side.

$m\angle RQS = \square$ — $180 - \square = \square$

The measure of $\square$ is 124°.

Check Your Progress Angles MNP and KNP are complementary. If $m\angle MNP = 23°$, find $m\angle KNP$.

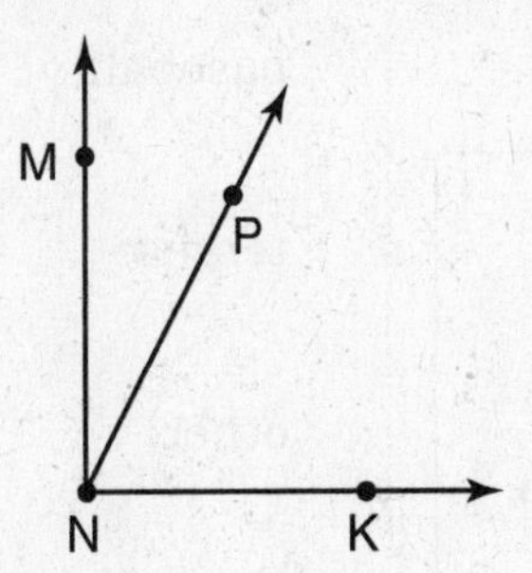

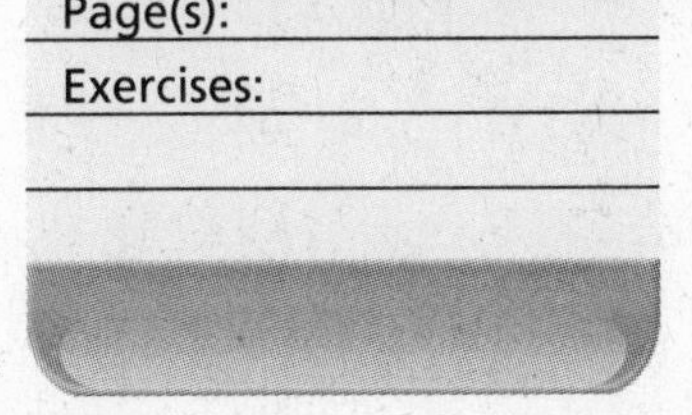

HOMEWORK ASSIGNMENT

Page(s):

Exercises:

10–3 Statistics: Display Data in a Circle Graph

MAIN IDEA

- Construct and interpret circle graphs.

BUILD YOUR VOCABULARY (pages 225–226)

A graph that shows data as parts of a ______ is a **circle graph**.

EXAMPLE Display Data in a Circle Graph

1 SPORTS In a survey, a group of middle school students were asked to name their favorite sport. The results are shown in the table. Make a circle graph of the data.

Sport	Percent
football	30%
basketball	25%
baseball	22%
tennis	8%
other	15%

- Find the degrees for each part. Round to the nearest whole degree.

football: ______ of 360° = 0.30 • 360° or ______

basketball: 25% of 360° = ______ • 360° or ______

baseball: ______ of 360° = 0.22 • 360° or about ______

tennis: 8% of 360° = ______ • 360° or about ______

other: ______ of 360° = 0.15 • 360° or about 

- Draw a circle with a radius marked as shown. Then use a ______ to draw the first angle, in this case ______. Repeat this step for each section.

WRITE IT

Write a proportion to convert 65% to the number of degrees in a part of a circle graph.

______

______

______

______

______

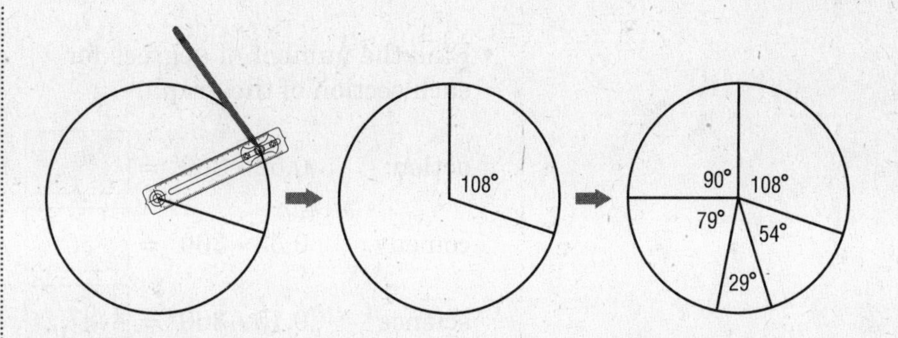

- Label each section of the graph with the category and ______. Give the graph a ______.

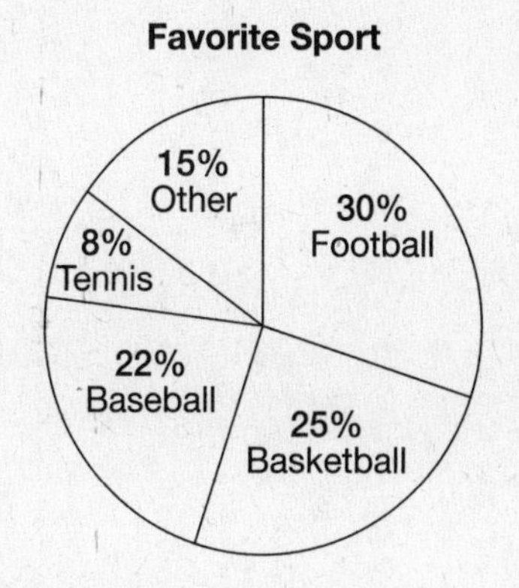

REVIEW IT

Explain how to convert a fraction to a decimal. *(Lesson 4-5)*

______

______

______

______

______

______

EXAMPLE **Construct a Circle Graph**

2 MOVIES **Gina has the following types of movies in her DVD collection. Make a circle graph of the data.**

Type of Movie	Numbers
action	24
comedy	15
science fiction	7

- Find the total number of DVDs: 24 + 15 + 7 or ______.
- Find the ______ that compares each number with the ______. Write the ratio as a ______ number rounded to the nearest hundredth.

action: ______ ≈ 0.52

comedy: ______ ≈ 0.33

science fiction: ______ ≈ 0.15

(continued on the next page)

- Find the number of degrees for each section of the graph.

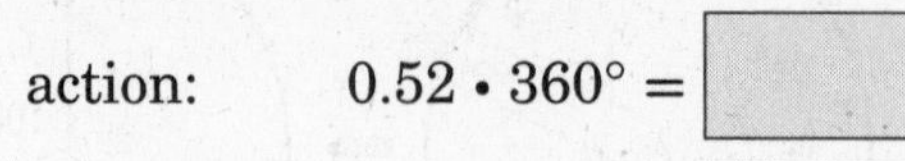

action: $0.52 \cdot 360° =$ ______

comedy: $0.33 \cdot 360° =$ ______

science fiction: $0.15 \cdot 360° =$ ______

Gina's DVD Collection

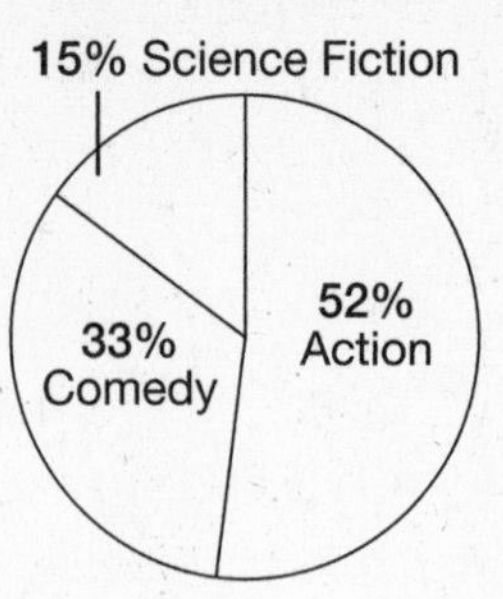

- Draw the circle graph.

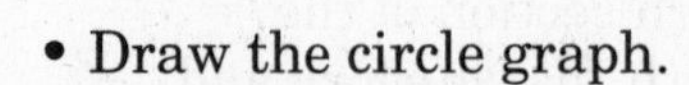

Check Your Progress

a. ICE CREAM In a survey, a group of students were asked to name their favorite flavor of ice cream. The results are shown in the table. Make a circle graph of the data.

Flavor	Percent
chocolate	30%
cookie dough	25%
peanut butter	15%
strawberry	10%
other	20%

b. MARBLES Michael has the following colors of marbles in his marble collection. Make a circle graph of the data.

Color	Number
black	12
green	9
red	5
gold	3

EXAMPLES Analyze a Circle Graph

VOTING The circle graph below shows the percent of voters in a town who are registered with a political party.

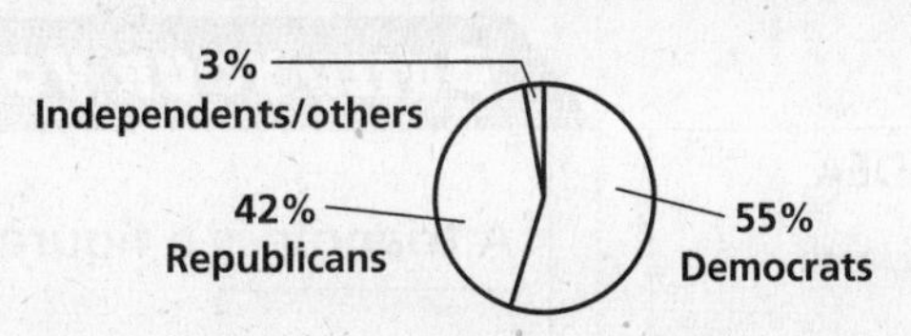

3 Which party has the most registered voters?

The largest section of the circle is the one representing ______. So, the Democratic party has the most registered voters.

4 If the town has 3,400 registered Republicans, about how many voters are registered in all?

Republicans: 42% of registered voters = ______

$$0.42 \times n = 3{,}400$$

$$0.42n = 3{,}400$$

$$n \approx 8{,}095$$

So, there are about ______ registered voters in all.

Check Your Progress **SPORTS The circle graph below shows the responses of middle school students to the question "Should teens be allowed to play professional sports?"**

a. Which response was the greatest?

Should Teens Be Allowed to Play Professional Sports?

Yes 55%

No 32%

No Opinion 13%

b. If there were 1,500 middle school students, how many had no opinion?

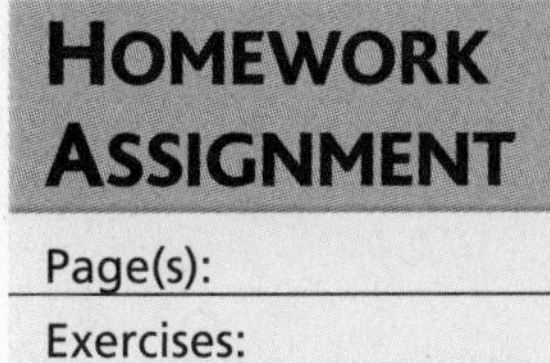

10–4 Triangles

MAIN IDEA

- Identify and classify triangles.

BUILD YOUR VOCABULARY (pages 225–226)

A **triangle** is a figure with three ______ and three ______.

Sides with the same ______ are **congruent segments**.

KEY CONCEPT

Angles of a Triangle The sum of the measures of the angles of a triangle is 180°.

FOLDABLES Record this relationship in your Foldable. Be sure to include an example.

EXAMPLE Find a Missing Measure

1 ALGEBRA Find $m\angle A$ in $\triangle ABC$ if $m\angle A = m\angle B$, and $m\angle C = 80°$.

Since the sum of the angle measures in a triangle is 180°,

$m\angle A + m\angle B + m\angle C =$ ______.

Let x represent $m\angle A$. Since $m\angle A = m\angle B$, x also represents

.

$x + x + 80 = 180$	Write the equation.
______ $+ 80 = 180$	$x + x = 2x$
$-$ ______ $\quad -$ ______	Subtract ______ from each side.
$\dfrac{2x}{\_\_\_} = \dfrac{100}{\_\_\_}$	Divide each side by 2.
$x =$ ______	So, $m\angle A =$ ______.

Check Your Progress **ALGEBRA** Find $m\angle M$ in $\triangle MNO$ if $m\angle N = 75°$ and $m\angle O = 67°$.

BUILD YOUR VOCABULARY (pages 225–226)

An **acute triangle** has all acute angles. A **right triangle** has one right angle. An **obtuse triangle** has one obtuse angle. A **scalene triangle** has no congruent sides. An **isosceles triangle** has at least 2 congruent sides. An **equilateral triangle** has three congruent sides.

EXAMPLE

2 TEST EXAMPLE An airplane has wings that are shaped like triangles. What is the missing measure of the angle?

A 41° **B** 31° **C** 26° **D** 21°

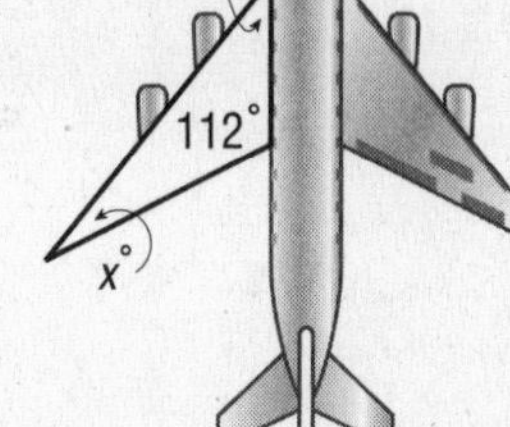

Read the Item

To find the missing measure, write and solve an equation.

Solve the Item

$x + \square + \square = 180$	The sum of the measures is 180.
$x + \square = 180$	Simplify.
$-159 \quad -159$	Subtract 159 from each side.
$x = \square$	

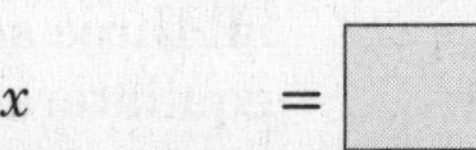

The missing measure is 21°. The answer is D.

Check Your Progress

MULTIPLE CHOICE A piece of fabric is shaped like a triangle. Find the missing angle measure.

F 73° **G** 49°

H 58° **J** 53°

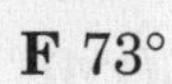

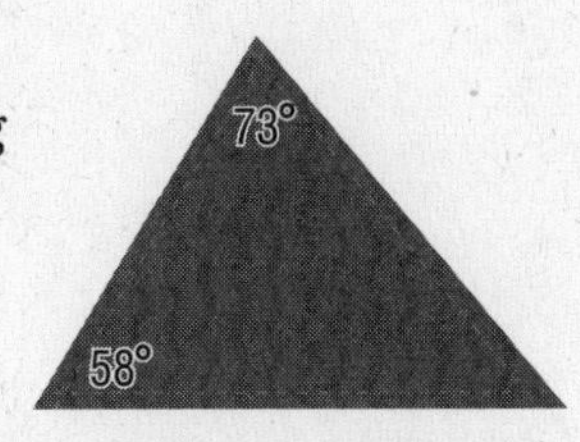

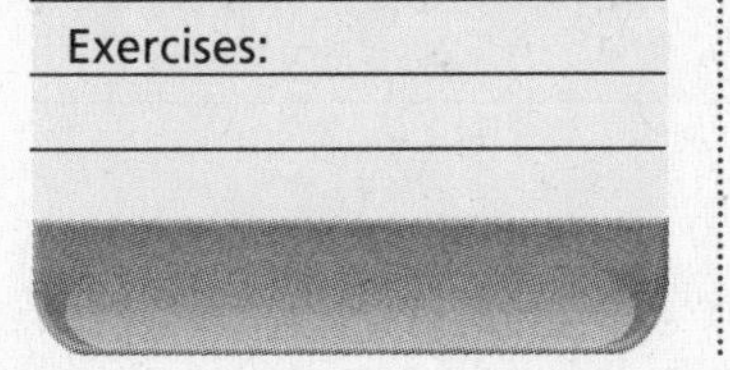
HOMEWORK ASSIGNMENT

Page(s):

Exercises:

10–5 Problem-Solving Investigation: Use Logical Reasoning

EXAMPLE Solve Using Logical Reasoning

MAIN IDEA

- Solve problems by using logical reasoning.

GEOMETRY Draw an equilateral triangle. How can you confirm that it is equilateral?

UNDERSTAND You know that equilateral triangles have ______ congruent sides. You need to confirm whether or not a drawn triangle is equilateral.

PLAN Draw an equilateral triangle. Measure the sides to confirm that all three sides are ______.

SOLVE Draw the triangle.

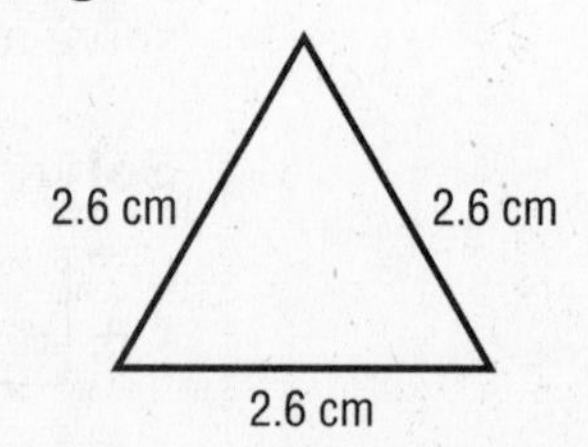

Measure the sides using a ruler or centimeter ruler. The side lengths are 2.6 centimeters, 2.6 centimeters, and 2.6 centimeters. Since all three sides are congruent, the triangle is equilateral.

CHECK Since all three sides are congruent, the triangle is equilateral. You can have someone else also measure the sides to check that the triangle is ______.

Check Your Progress **GEOMETRY** Do the angles in an equilateral triangle have a special relationship?

HOMEWORK ASSIGNMENT

Page(s):

Exercises:

10–6 Quadrilaterals

MAIN IDEA

- Identify and classify quadrilaterals.

BUILD YOUR VOCABULARY (pages 225–226)

A **quadrilateral** is a ______ figure with ______ sides and four ______.

A **parallelogram** is a quadrilateral with opposite sides ______ and opposite sides ______.

A **trapezoid** is a ______ with one pair of ______ sides.

A **rhombus** is a parallelogram with four congruent sides.

FOLDABLES

ORGANIZE IT

Record what you learn about quadrilaterals. Illustrate and describe the five types of quadrilaterals discussed in this chapter.

What I Know About Polygons	What I Need to Know	What I've Learned

EXAMPLES Classify Quadrilaterals

Classify the quadrilateral using the name that best describes it.

1

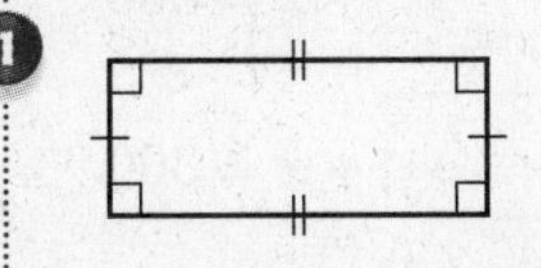

The quadrilateral has 4 ______ angles and opposite sides are ______. It is a ______.

2

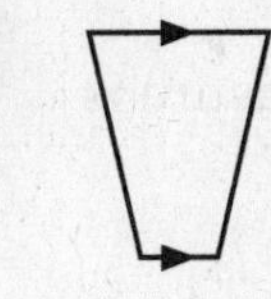

The quadrilateral has ______ pair of ______ sides.

It is a ______.

KEY CONCEPT

Angles of a Quadrilateral The sum of the measures of the angles of a quadrilateral is 360°.

Check Your Progress **Classify the quadrilateral using the name that *best* describes it.**

a.

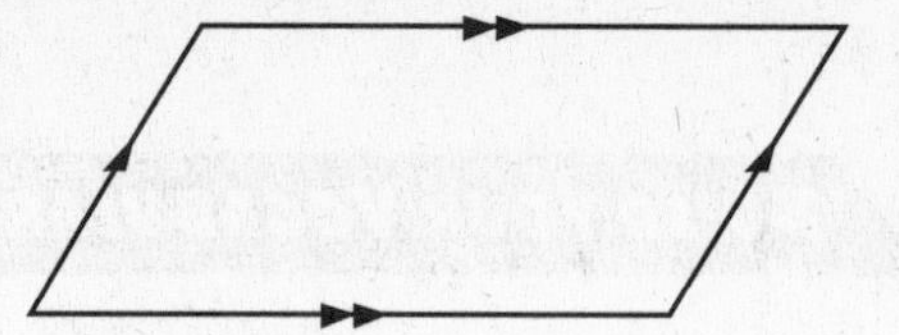

b.

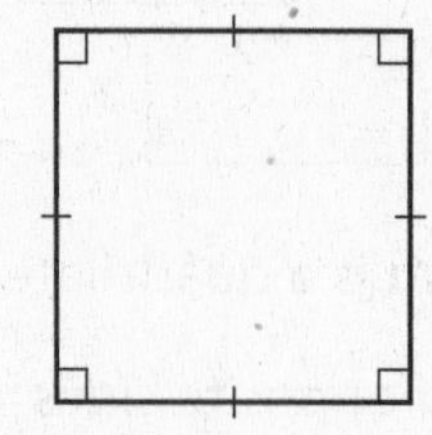

EXAMPLE Find a Missing Measure

3 ALGEBRA **Find the value of x in the quadrilateral shown.**

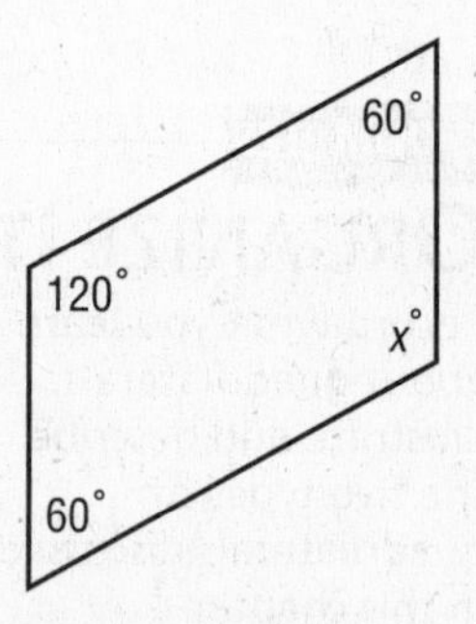

Write and solve an equation. Let x represent the missing measure.

$\square + \square + \square + x = 360$ The sum of the measures is 360°.

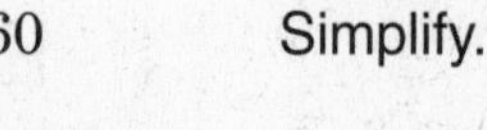

$\square + x = 360$ Simplify.

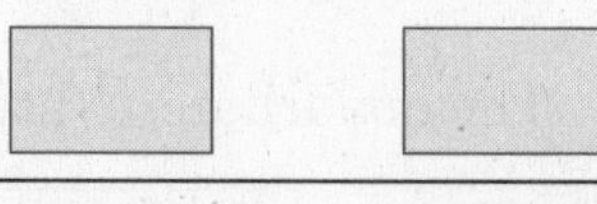

Subtract $\square$ from both sides.

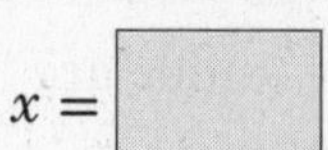

$x = \square$

So, the missing angle measure is $\square$.

Check Your Progress Find the missing angle measure in the quadrilateral.

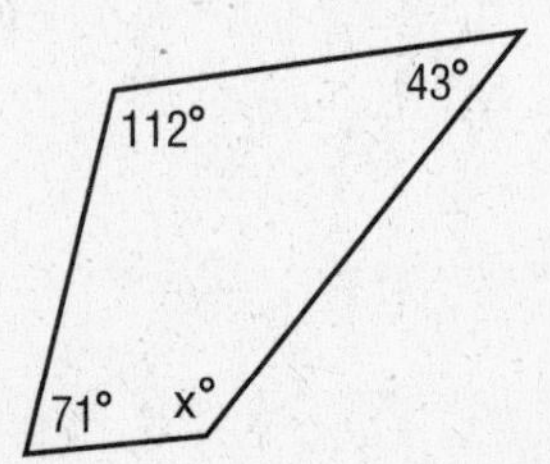

HOMEWORK ASSIGNMENT

Page(s):

Exercises:

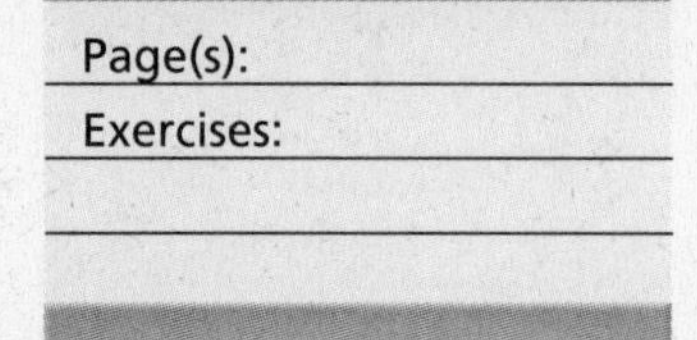

10–7 Similar Figures

MAIN IDEA

- Determine whether figures are similar and find a missing length in a pair of similar figures.

KEY CONCEPT

Similar Figures If two figures are similar, then

- the corresponding sides are proportional, and
- the corresponding angles are congruent.

BUILD YOUR VOCABULARY (pages 225–226)

Figures that have the same ______ but not necessarily the same ______ are **similar figures**.

The ______ of similar figures that "match" are **corresponding sides**.

The ______ of similar figures that "match" are **corresponding angles**.

EXAMPLE Identify Similar Figures

1 **Which rectangle below is similar to rectangle *FGHI*?**

F 9 ft G, 3 ft, I H

L 8 ft M, 2 ft, O N

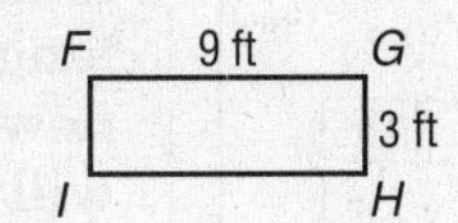

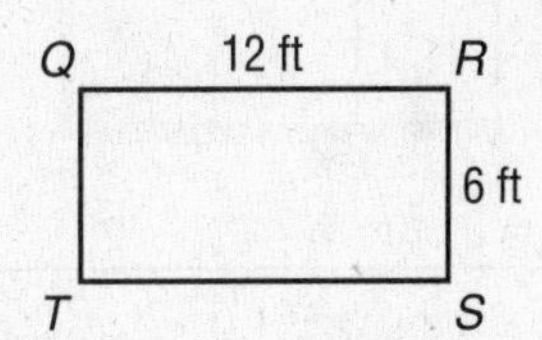

Compare the ratios of the corresponding sides.

Rectangle *LMNO*	Rectangle *ABCD*	Rectangle *QRST*
$\frac{FG}{LM} = \frac{9}{8}$	$\frac{FG}{AB} =$ ______	$\frac{FG}{QR} = \frac{9}{12}$
$\frac{GH}{MN} =$ ______	$\frac{GH}{BC} = \frac{3}{2}$	$\frac{GH}{RS} =$ ______
______	______	______

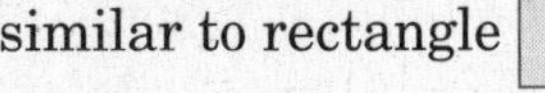

So, rectangle *FGHI* is similar to rectangle ______.

Check Your Progress Which rectangle from Example 1 is similar to rectangle *WXYZ* shown?

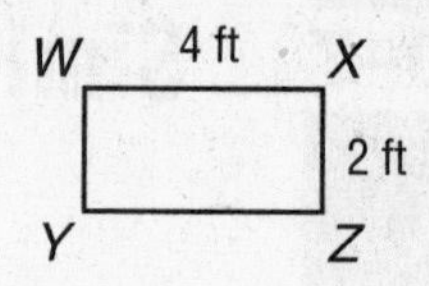

BUILD YOUR VOCABULARY (pages 225–226)

Indirect measurement uses similar figures to find the length, width, or height of objects that are too difficult to measure directly.

FOLDABLES

ORGANIZE IT

Use your Foldable to record what you learn about similar figures and indirect measurement.

What I Know About Polygons	What I Need to Know	What I've Learned

EXAMPLE

2 ARCHITECTURE A rectangular picture window 12 feet long and 6 feet wide needs to be shortened to 9 feet in length to fit a redesigned wall. If the architect wants the new window to be similar to the old window, how wide will the new window be?

$\frac{12}{9} = \frac{6}{w}$ Write a proportion.

$12w =$ ______ Find the cross products.

$12w =$ ______ Simplify.

$w =$ ______ Divide each side by ______.

So, the width of the new window will be ______ feet.

Check Your Progress Tom has a rectangular garden that has a length of 12 feet and a width of 8 feet. He wishes to start a second garden that is similar to the first and will have a width of 6 feet. Find the length of the new garden.

HOMEWORK ASSIGNMENT

Page(s):

Exercises:

10–8 Polygons and Tessellations

MAIN IDEA

- Classify polygons and determine which polygons can form a tessellation.

BUILD YOUR VOCABULARY (pages 225–226)

A **polygon** is a simple, closed figure formed by three or more straight line segments.

A **regular polygon** has all sides congruent and all angles congruent.

A polygon is named by the number of sides it has: **pentagon** (5 sides), **hexagon** (6 sides), **heptagon** (7 sides), **octagon** (8 sides), **nonagon** (9 sides), and **decagon** (10 sides).

EXAMPLES Classify Polygons

Determine whether each figure is a polygon.

1 The figure is not a polygon since it has a ______ side.

2 This figure has 6 sides that are not all of equal length. It is a ______ that is not ______.

Check Your Progress **Determine whether each figure is a polygon. If it is, classify the polygon and state whether it is regular. If it is not a polygon, explain why.**

a.

b.

FOLDABLES

ORGANIZE IT

Use your Foldable to record what you learn about polygons and tessellations. Explain how a tessellation can be made with several kinds of polygons.

What I Know About Polygons	What I Need to Know	What I've Learned

BUILD YOUR VOCABULARY (pages 225–226)

A repetitive pattern of polygons that fit together with no ______ or ______ is called a **tessellation**.

EXAMPLE Tessellations

3 **PATTERNS Ms. Pena is creating a pattern on her wall. She wants to use regular hexagons. Can Ms. Pena make a tessellation with regular hexagons?**

The measure of each angle in a regular hexagon is ______.

The sum of the measures of the angles where the vertices meet must be 360°.

So, solve $120n = 360$.

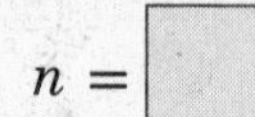 Write the equation.

$\frac{120n}{120} = \frac{360}{120}$ Divide each side by ______.

$n =$ ______

Since 120° divides evenly into 360°, the sum of the measures where the vertices meet is ______. So, Ms. Pena can make a tessellation with regular hexagons.

Check Your Progress **QUILTING** Emily is making a quilt using fabric pieces shaped as equilateral triangles. Can Emily tessellate the quilt with these fabric pieces?

HOMEWORK ASSIGNMENT

Page(s):

Exercises:

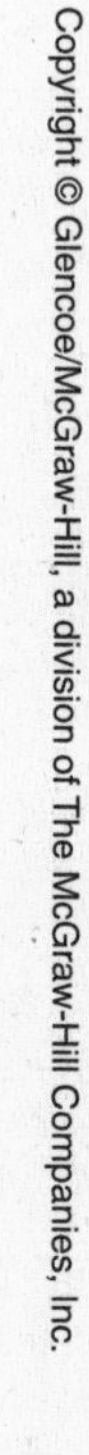

10-9 Translations

Main Idea

- Graph translations of polygons on a coordinate plane.

BUILD YOUR VOCABULARY (pages 225–226)

A **transformation** maps one figure onto another.

A **translation** is a transformation where a figure is moved without turning it.

The original figure and the translated figure are **congruent figures**.

EXAMPLE Graph a Translation

1 Translate $\triangle ABC$ 5 units left and 1 unit up.

- Move each vertex of the figure 5 units left and 1 unit up. Label the new vertices A', B', and C'.
- Connect the vertices to draw the triangle. The coordinates of the vertices of the new figure are ______, ______, and ______.

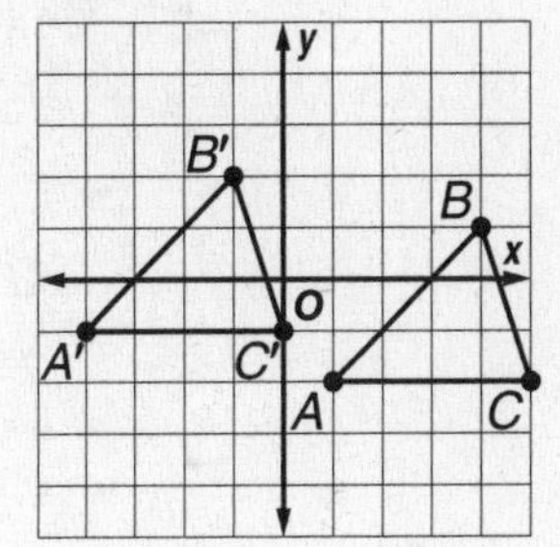

Check Your Progress

Translate $\triangle DEF$ 3 units left and 2 units down.

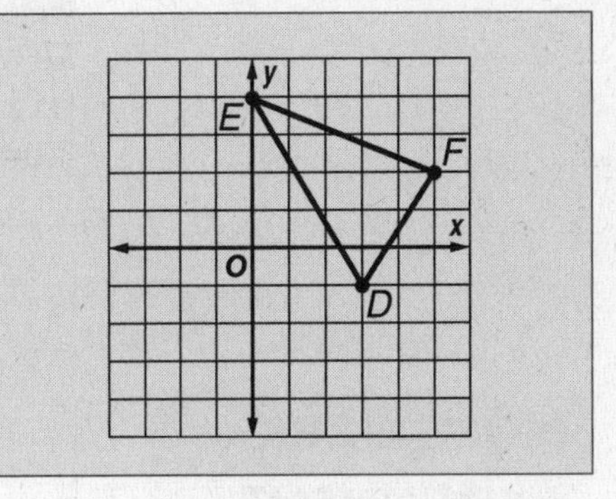

Remember It

The order of a translation of a figure does not matter. Moving a figure to the side x units and then up y units is the same as moving it up y units and then to the side x units.

EXAMPLE Find Coordinates of a Translation

2 **Trapezoid *GHIJ* has vertices *G*(−4, 1), *H*(−4, 3,), *I*(−2, 3), and *J*(−1, 1). Find the vertices of trapezoid *G′H′I′J′* after a translation of 5 units right and 3 units down. Then graph the figure and its translated image.**

Add ______ to each x-coordinate. Add ______ to each y-coordinate.

Vertices of Trapezoid *GHIJ*	$(x + 5, y - 3)$	Vertices of Trapezoid *G′H′I′J′*
$G(-4, 1)$		$G'(1, -2)$
$H(-4, 3)$	$(-4 + 5, 3 - 3)$	
	$(-2 + 5, 3 - 3)$	
$J(-1, 1)$		$J'(4, -2)$

The coordinates of trapezoid *G′H′I′J′* are *G′* ______, *H′* ______, *I′* ______, and *J′* ______.

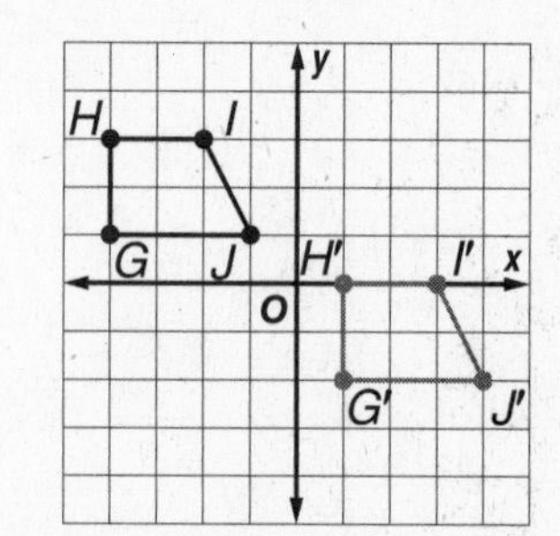

Check Your Progress Triangle *MNO* has vertices *M*(−5, −3), *N*(−7, 0), and *O*(−2, 3). Find the vertices of triangle *M′N′O′* after a translation of 6 units right and 3 units up. Then graph the figure and its translated image.

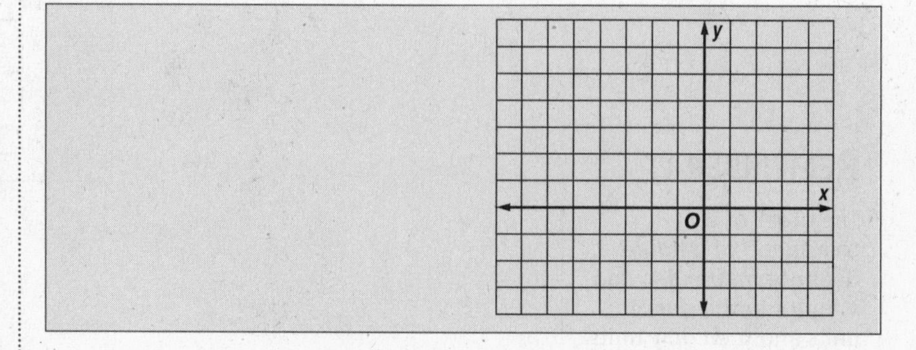

HOMEWORK ASSIGNMENT

Page(s):

Exercises:

10–10 Reflections

MAIN IDEA

- Identify figures with line symmetry and graph reflections on a coordinate plane.

BUILD YOUR VOCABULARY (pages 225–226)

Figures that ______ exactly when they are folded in ______ have **line symmetry**.

Each fold line is called a **line of symmetry**.

EXAMPLES Identify Lines of Symmetry

LETTERS **Determine whether each letter has a line of symmetry. If so, copy the figure and draw all lines of symmetry.**

1. X — This figure has line ______.
   There are ______ lines of symmetry.

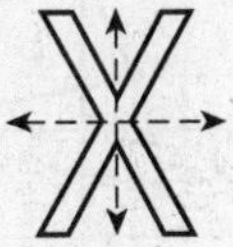

2. Y — This figure has line symmetry.
   There is ______ line of symmetry.

3. P — This figure ______ have line symmetry.

Check Your Progress **Determine whether each figure has line symmetry. If so, copy the figure and draw all lines of symmetry.**

a.

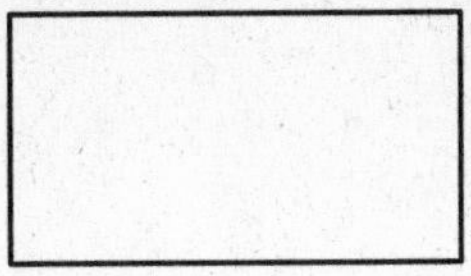

b.

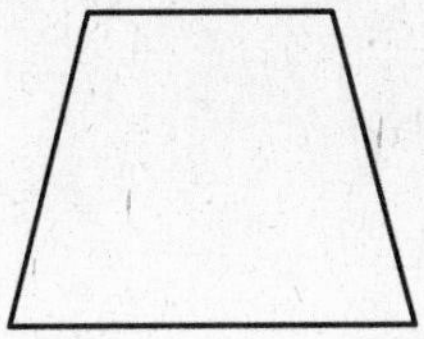

REMEMBER IT

Vertices of a figure receive double prime symbols (″) after they have been transformed twice.

BUILD YOUR VOCABULARY (pages 225–226)

A **reflection** is a mirror ________ of the original figure that is the result of a transformation over a ________ called a **line of reflection.**

EXAMPLE Reflect a Figure Over the x-axis

4 Quadrilateral $QRST$ has vertices $Q(-1, 1)$, $R(0, 3)$, $S(3, 2)$, and $T(4, 0)$. Graph the figure and its reflected image over the x-axis. Then find the coordinates of the reflected image.

The x-axis is the line of reflection. So, plot each vertex of $Q'R'S'T'$ the same distance from the x-axis as its corresponding vertex on $QRST$.

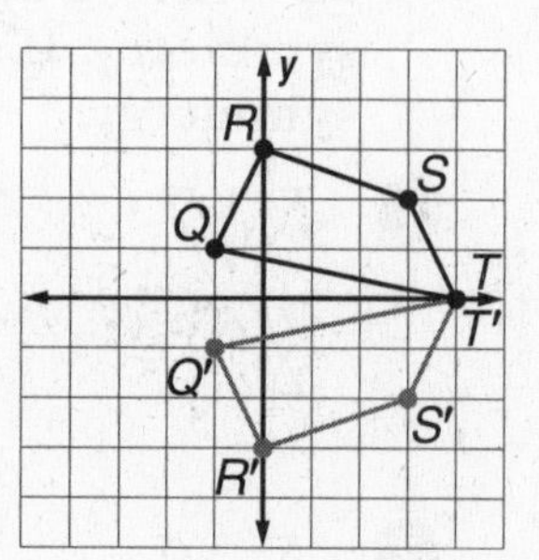

Q' ________ R' ________

S' ________ T' ________

Check Your Progress Quadrilateral $ABCD$ has vertices $A(-3, 2,)$ $B(-1, 5)$, $C(3, 3)$, and $D(2, 1)$. Graph the figure and its reflection over the x-axis. Then find the coordinates of the reflected image.

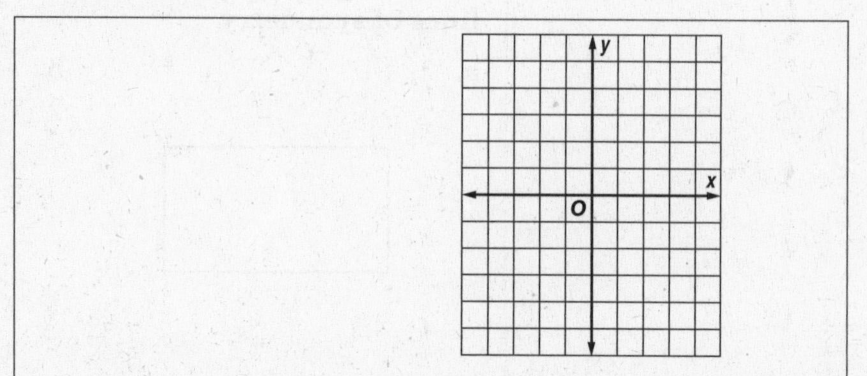

EXAMPLE **Reflect a Figure over the *y*-axis**

5 Triangle *XYZ* has vertices $X(1, 2)$, $Y(2, 1)$, and $Z(1, -2)$. Graph the figure and its reflected image over the *y*-axis. Then find the coordinates of the reflected image.

The *y*-axis is the line of reflection. So, plot each vertex of $X'Y'Z'$ the same distance from the *y*-axis and its corresponding vertex on *XYZ*.

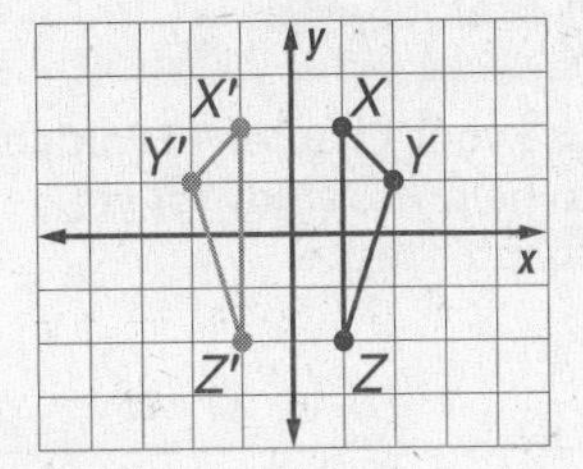

Check Your Progress Triangle *QRS* has vertices $Q(3, 4)$, $R(1, 0)$, and $S(6, 2)$. Graph the figure and its reflection over the *y*-axis. Then find the coordinates of the reflected image.

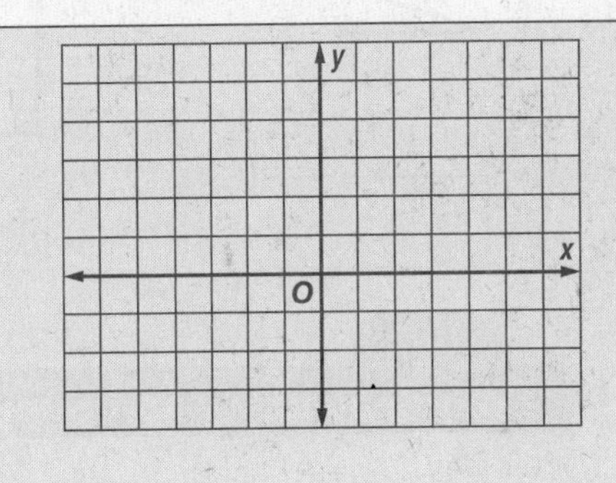

HOMEWORK ASSIGNMENT

Page(s):

Exercises:

CHAPTER 10

BRINGING IT ALL TOGETHER

STUDY GUIDE

FOLDABLES	VOCABULARY PUZZLEMAKER	BUILD YOUR VOCABULARY
Use your **Chapter 10 Foldable** to help you study for your chapter test.	To make a crossword puzzle, word search, or jumble puzzle of the vocabulary words in Chapter 10, go to: glencoe.com	You can use your completed **Vocabulary Builder** (*pages 225–226*) to help you solve the puzzle.

10-1 Angle Relationships

Classify each angle as *acute*, *obtuse*, or *right*.

1.

2.

3. 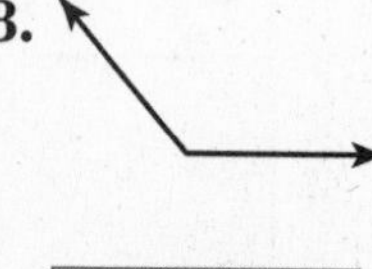

10-2 Complementary and Supplementary Angles

Complete each sentence.

4. The sum of the measures of ______ angles is 180°.

5. The sum of the measures of ______ angles is 90°.

6. If $\angle A$ and $\angle B$ are supplementary angles and $m\angle B = 43°$, find $m\angle A$.

10-3 Statistics: Display Data in a Circle Graph

Find the number of degrees for each part of the graph at the right.

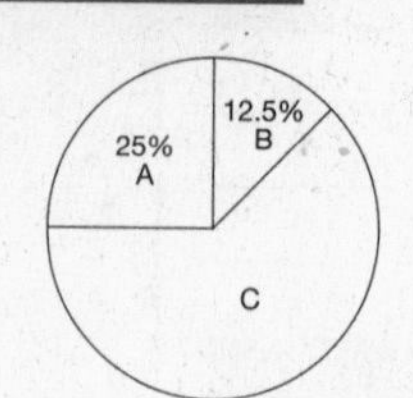

7. A ______ 8. B ______ 9. C ______

10-4 Triangles

Complete the table to help you remember the ways to classify triangles.

	Type of Triangle	Classified by Angles or Sides	Description
10.	acute	angles	
11.	obtuse		
12.		sides	no congruent sides
13.			1 right angle
14.	equilateral		

10-5 Problem-Solving Investigation: Logical Reasoning

15. RACES Marcus, Elena, Pedro, Keith, and Darcy ran a 2-mile race. Darcy finished directly after Pedro, Elena finished before Marcus, and Keith finished first. If Pedro finished third, order the runners from first to last.

10-6 Quadrilaterals

Find the value of x in the quadrilateral.

16.

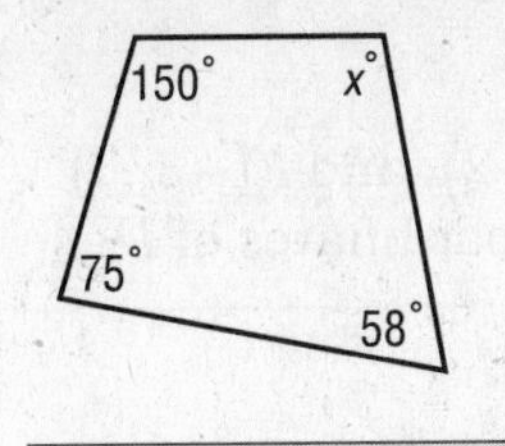

17.

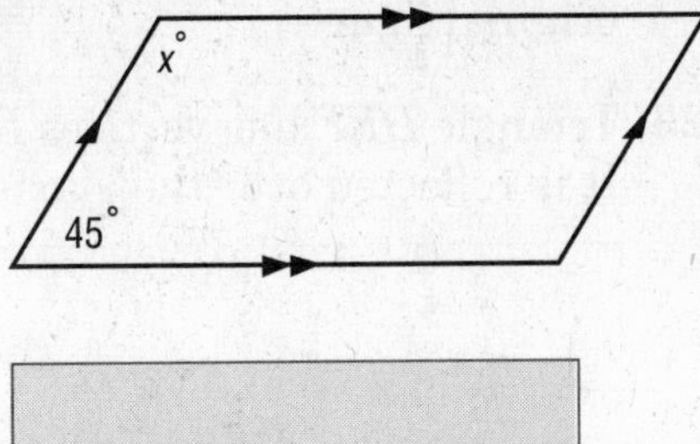

10-7 Similar Figures

18. Find the value of x if $\triangle ABC \sim \triangle DEF$.

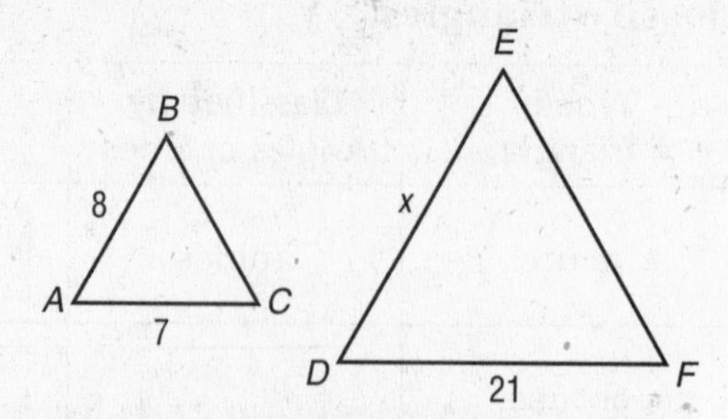

10-8 Polygons and Tessellations

Underline the correct term to complete each sentence.

19. A polygon can have (two, three) or more straight lines.

20. To find the sum of the angle measures in a regular polygon, draw all the diagonals from one vertex, count the number of (angles, triangles) formed, and multiply by 180°.

10-9 Translations

21. Triangle ABC with vertices $A(2, 4)$, $B(-4, 6)$, and $C(1, -5)$ is translated 2 units right and 3 units down. What are the coordinates of B?

10-10 Reflections

Underline the correct word(s) to complete the sentence.

22. The image of a reflection is (larger than, the same size as) the original figure.

23. Triangle DEF has vertices $D(-5, 2)$, $E(-4, -2)$, and $F(-3, 0)$. It is reflected over the y-axis. What are the coordinates of D?

ARE YOU READY FOR THE CHAPTER TEST?

Math Online

Visit **glencoe.com** to access your textbook, more examples, self-check quizzes, and practice tests to help you study the concepts in Chapter 10.

Check the one that applies. Suggestions to help you study are given with each item.

☐ **I completed the review of all or most lessons without using my notes or asking for help.**

- You are probably ready for the Chapter Test.
- You may want to take the Chapter 10 Practice Test on page 567 of your textbook as a final check.

☐ **I used my Foldable or Study Notebook to complete the review of all or most lessons.**

- You should complete the Chapter 10 Study Guide and Review on pages 563–566 of your textbook.
- If you are unsure of any concepts or skills, refer to the specific lesson(s).
- You may want to take the Chapter 10 Practice Test on page 567 of your textbook.

☐ **I asked for help from someone else to complete the review of all or most lessons.**

- You should review the examples and concepts in your Study Notebook and Chapter 10 Foldable.
- Then complete the Chapter 10 Study Guide and Review on pages 563–566 of your textbook.
- If you are unsure of any concepts or skills, refer to the specific lesson(s).
- You may also want to take the Chapter 10 Practice Test on page 567 of your textbook.

Student Signature

Parent/Guardian Signature

Teacher Signature

Measurement: Two- and Three-Dimensional Figures

Use the instructions below to make a Foldable to help you organize your notes as you study the chapter. You will see Foldable reminders in the margin of this Interactive Study Notebook to help you in taking notes.

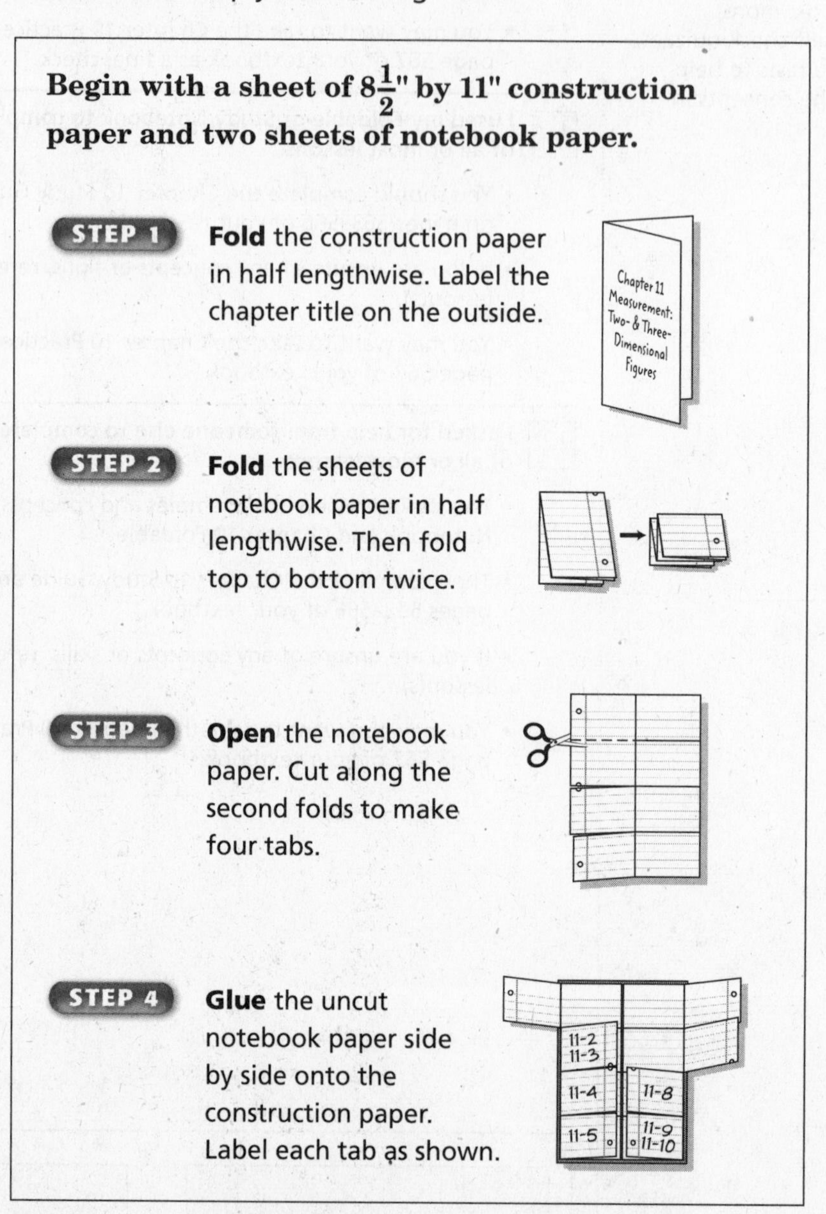

Begin with a sheet of $8\frac{1}{2}$" by 11" construction paper and two sheets of notebook paper.

STEP 1 **Fold** the construction paper in half lengthwise. Label the chapter title on the outside.

STEP 2 **Fold** the sheets of notebook paper in half lengthwise. Then fold top to bottom twice.

STEP 3 **Open** the notebook paper. Cut along the second folds to make four tabs.

STEP 4 **Glue** the uncut notebook paper side by side onto the construction paper. Label each tab as shown.

NOTE-TAKING TIP: When you take notes, it is helpful to write key vocabulary words, definitions, concepts, or procedures as clearly and concisely as possible.

BUILD YOUR VOCABULARY

This is an alphabetical list of new vocabulary terms you will learn in Chapter 11. As you complete the study notes for the chapter, you will see Build Your Vocabulary reminders to complete each term's definition or description on these pages. Remember to add the textbook page number in the second column for reference when you study.

Vocabulary Term	Found on Page	Definition	Description or Example
base			
circle			
circumference			
composite figure			
cone			
cylinder			
diameter			
edge			
face			

(continued on the next page)

Vocabulary Term	Found on Page	Definition	Description or Example
height			
lateral face			
prism			
pyramid			
radius			
rectangular prism			
solid			
sphere			
three-dimensional figure			
triangular prism			
vertex			
volume			

11-1 Area of Parallelograms

MAIN IDEA

- Find the areas of parallelograms.

KEY CONCEPT

Area of a Parallelogram The area *A* of a parallelogram equals the product of its base *b* and height *h*.

BUILD YOUR VOCABULARY (pages 255–256)

The **base** is any ______ of a parallelogram.

The **height** is the length of the segment ______ to the ______ with endpoints on ______ sides.

EXAMPLE Find the Area of a Parallelogram

1 Find the area of the parallelogram.

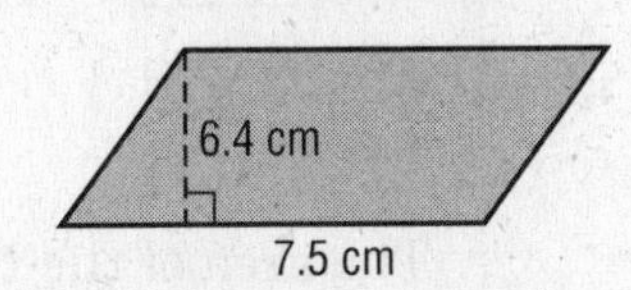

Estimate $A =$ ______ $\cdot$ ______ or ______ cm^2

$A = bh$ Area of a parallelogram

$A =$ ______ $\cdot$ ______ Replace ______ with 7.5 and ______ with 6.4.

$A =$ ______ Multiply.

The area of the parallelogram is ______ square centimeters. This is the same as the estimate.

Check Your Progress Find the area of the parallelogram.

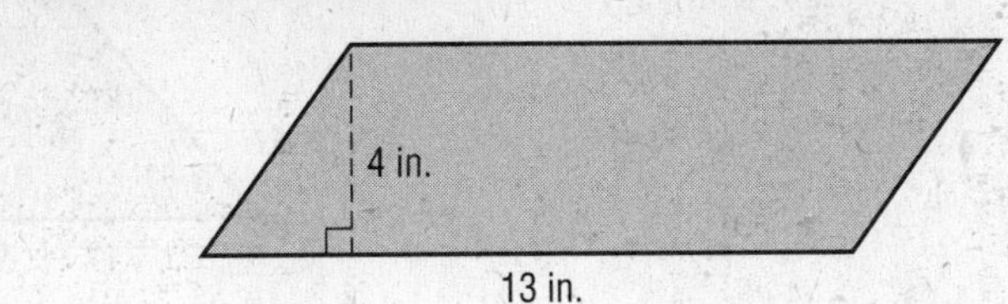

HOMEWORK ASSIGNMENT

Page(s):

Exercises:

11–2 Areas of Triangles and Trapezoids

MAIN IDEA

- Find the areas of triangles and trapezoids.

KEY CONCEPT

Area of a Triangle The area *A* of a triangle equals half the product of its base *b* and height *h*.

EXAMPLE Find the Area of a Triangle

1 Find the area of the triangle below.

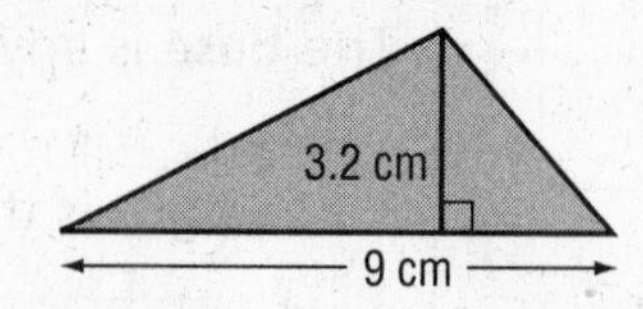

Estimate $\frac{1}{2}(9)(3) =$ ______

$A = \frac{1}{2}bh$ Area of a triangle.

$A = \frac{1}{2}$ ______ ______ Replace *b* with ______ and *h* with ______.

$A =$ ______ Multiply.

The area of the triangle is 14.4 ______.

This is close to the estimate.

Check Your Progress Find the area of the triangle below.

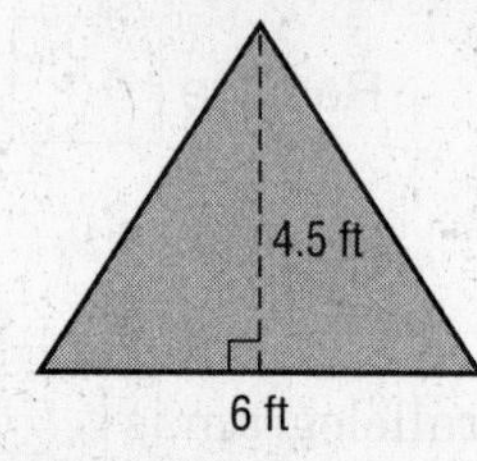

EXAMPLE Find the Area of a Trapezoid

2 Find the area of the trapezoid below.

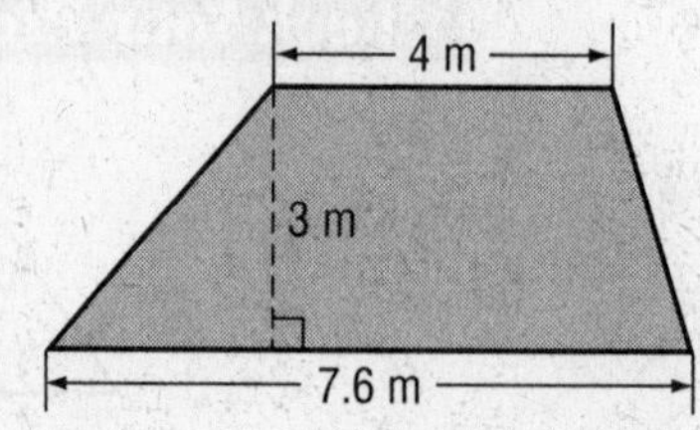

The bases are ______ meters and ______ meters.

The height is ______ meters.

KEY CONCEPT

Area of a Trapezoid The area A of a trapezoid equals half the product of the height h and the sum of the bases b_1 and b_2.

$A = \frac{1}{2}h(b_1 + b_2)$ Area of a trapezoid

$A = \frac{1}{2}(3)$ ______ Replace h with ______, b_1 with ______, and b_2 with ______.

$A = \frac{1}{2}$ ______ (11.6) Add ______ and ______.

$A =$ ______ Multiply.

The area of the trapezoid is ______ square meters.

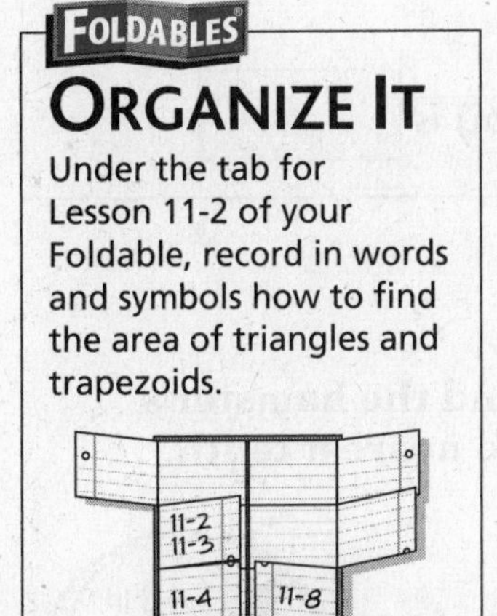

FOLDABLES

ORGANIZE IT

Under the tab for Lesson 11-2 of your Foldable, record in words and symbols how to find the area of triangles and trapezoids.

Check Your Progress Find the area of the trapezoid below.

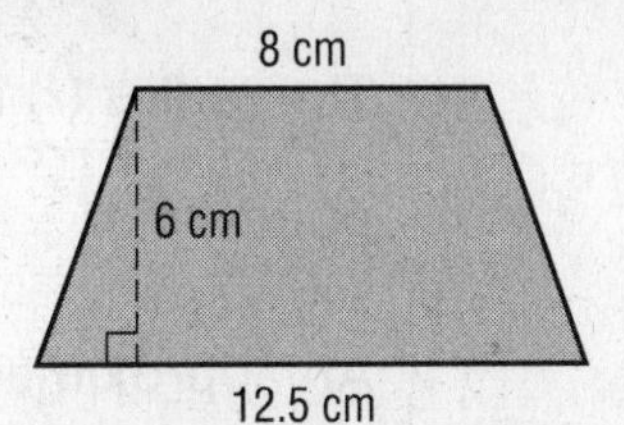

HOMEWORK ASSIGNMENT

Page(s):

Exercises:

11-3 Circles and Circumference

MAIN IDEA

- Find the circumference of circles.

BUILD YOUR VOCABULARY (pages 255–256)

A **circle** is a set of all points in a plane that are the ______ distance from a given ______ called the **center**.

The **diameter** (d) is the distance ______ a ______ through its center.

The **circumference** (C) is the distance ______ a circle.

The **radius** (r) is the distance from the ______ to any point on a ______.

An approximation often used for π **(pi)** is ______.

KEY CONCEPT

Circumference of a Circle The circumference C of a circle is equal to its diameter d times π, or 2 times its radius r times π.

EXAMPLE Find Circumference

1 PETS **Find the circumference around the hamster's running wheel shown. Round to the nearest tenth.**

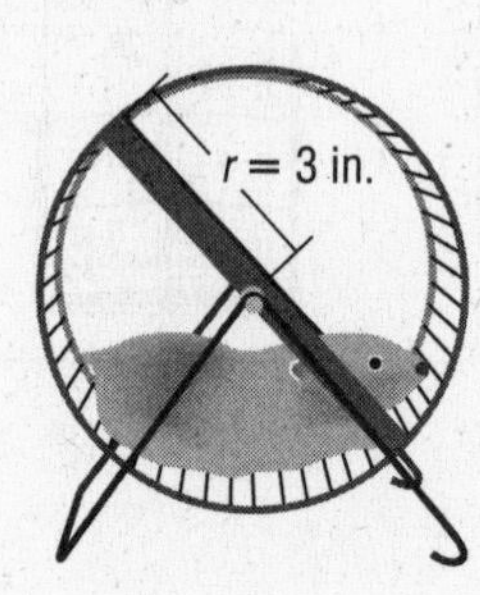

$C = 2\pi r$

$C = 2$ ______ (3)

$C =$ ______ Multiply.

The circumference is about ______ inches.

REMEMBER IT

All circumferences are estimates since 3.14 is an estimated value of pi.

Check Your Progress **SWIMMING POOL**

A new children's swimming pool is being built at the local recreation center. The pool is circular in shape with a diameter of 18 feet. Find the circumference of the pool. Round to the nearest tenth.

EXAMPLE **Find Circumference**

2 Find the circumference of a circle with a diameter of 49 centimeters.

Since 49 is a multiple of 7, use ______ for π.

$C = \pi d$ — Circumference of a circle

$C \approx \frac{22}{7} \cdot$ ______ — Replace ______ with $\frac{22}{7}$ and d with ______.

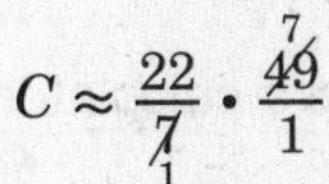

$C \approx \frac{22}{7} \cdot \frac{49}{1}$ — Divide by the , 7.

$C \approx$ ______ — Multiply.

The circumference is about 154 ______.

Check Your Progress Find the circumference of a circle with a radius of 35 feet.

HOMEWORK ASSIGNMENT

Page(s):

Exercises:

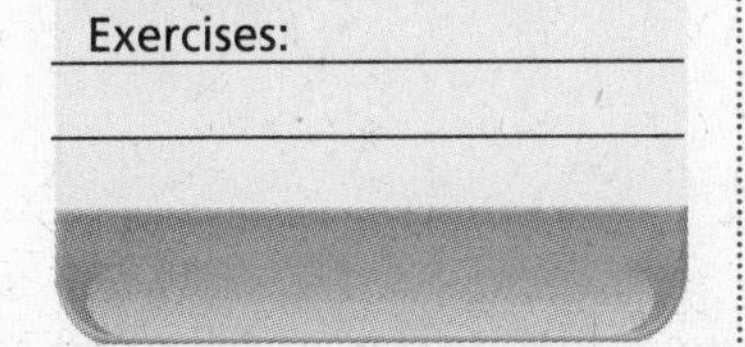

11-4 Area of Circles

EXAMPLES Find the Areas of Circles

MAIN IDEA

- Find the areas of circles.

KEY CONCEPT

Area of a Circle The area A of a circle equals the product of pi (π) and the square of its radius r.

1 Find the area of the circle at the right.

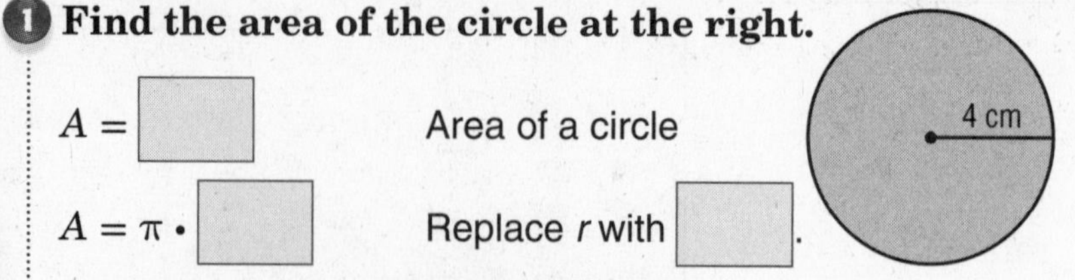

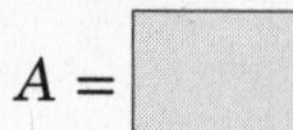

$A =$ Area of a circle

$A = \pi \cdot$ Replace r with .

π × 2 x^2 ENTER

The area of the circle is approximately square centimeters.

2 KOI Find the area of the koi pond shown.

The diameter of the pond is 3.6 meters, so the radius is $\frac{1}{2}(3.6)$ or 1.8 meters.

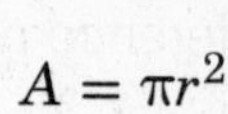

$A = \pi r^2$ Area of a circle

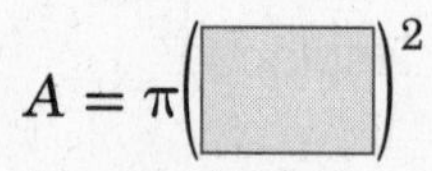

$A = \pi(\quad)^2$ Replace r with .

$A \approx$ Use a calculator.

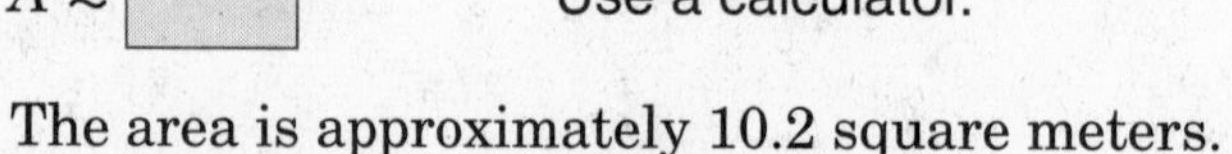

The area is approximately 10.2 square meters.

Check Your Progress

a. Find the area of the circle below.

b. COINS Find the area of a nickel with a diameter of 2.1 centimeters.

HOMEWORK ASSIGNMENT

Page(s):

Exercises:

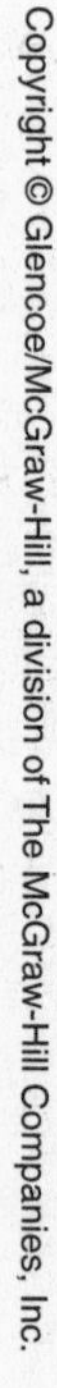

BUILD YOUR VOCABULARY (pages 255–256)

A **sector** of a circle is a region of a circle bounded by ________ radii.

EXAMPLE

3 TEST EXAMPLE Mr. McGowan made an apple pie with a diameter of 10 inches. He cut the pie into 6 equal slices. Find the approximate area of each slice.

A 3 in^2 **B** 13 in^2 **C** 16 in^2 **D** 52 in^2

Read the Item

You can use the diameter to find the total area of the pie and then divide that result by 6 to find the area of each slice.

Solve the Item

Find the area of the whole pie.

$A = \pi r^2$ Area of a circle

$A = \pi(\square)^2$ Replace r with ________.

$A \approx 78$ Multiply.

Find the area of one slice.

$78 \div \square = 13$

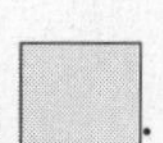

The area of each slice is approximately 13 square inches.

The correct answer is ________.

Check Your Progress **MULTIPLE CHOICE** The floor of a merry-go-round at the amusement park has a diameter of 40 feet. The floor is divided evenly into eight sections, each having a different color. Find the area of each section of the floor.

F 15.7 ft^2 **H** 62.8 ft^2

G 20 ft^2 **J** 157 ft^2

11–5

Problem-Solving Investigation: Solve a Simpler Problem

MAIN IDEA

- Solve problems by solving a simpler problem.

EXAMPLE Use the Solve a Simpler Problem Strategy

PAINT Ben and Shelia are going to paint the wall of a room as shown in the diagram. What is the area that will be painted?

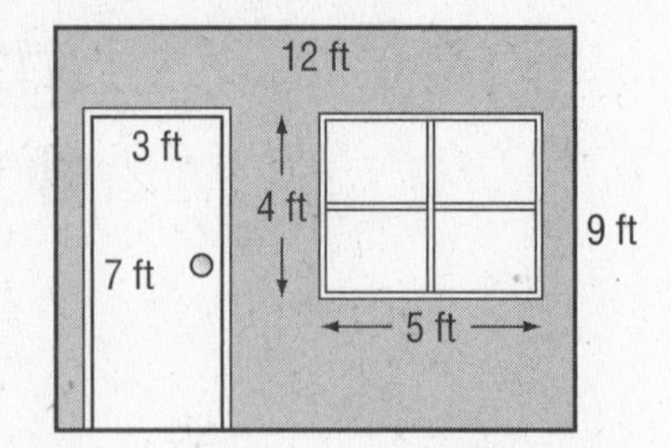

UNDERSTAND You know the dimensions of the wall including the door and window. You also know the dimensions of the door and window. You need to find the area of the wall not including the door and window.

PLAN Find the area of the wall including the door and window. Then subtract the area of the door and the window.

SOLVE area of wall including door and window:

$A = lw$

$A = 12 \cdot 9$ or ______ square feet

area of door:

$A = lw$

$A = 3 \cdot 7$ or ______ square feet

area of window:

$A = lw$

$A = 5 \cdot 4$ or ______ square feet

The total area to be painted is $108 - 21 - 20$ or ______ square feet.

CHECK The area to be painted is 67 square feet. Add the area of the door and the window. $67 + 21 + 20$ is 108 square feet. So, the answer is correct.

Check Your Progress Karen is placing a rectangular area rug measuring 8 feet by 10 feet in a rectangular dining room that measures 14 feet by 18 feet. Find the area of the flooring that is not covered by the area rug.

HOMEWORK ASSIGNMENT

Page(s):

Exercises:

11-6 Area of Composite Figures

MAIN IDEA

- Find the areas of composite figures.

BUILD YOUR VOCABULARY (pages 255–256)

A **composite figure** is made of **triangles**, quadrilaterals, **semicircles**, and other ________ figures.

A **semicircle** is ________ of a circle.

EXAMPLE Find the Area of a Composite Figure

1 Find the area of the figure in square centimeters.

The figure can be separated into a ________ and a ________. Find the area of each.

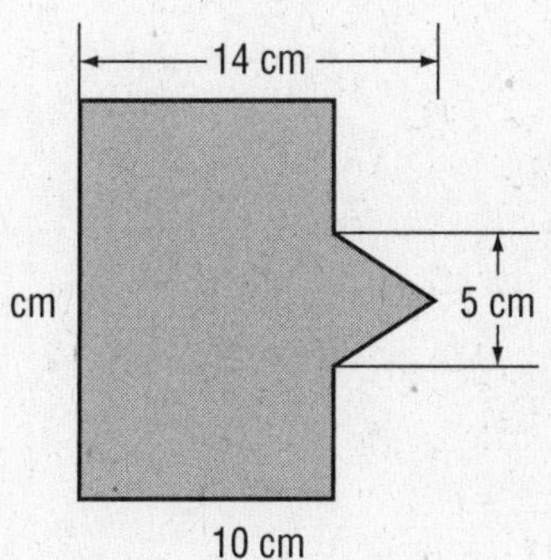

Area of Rectangle

$A = \ell w$

$A = 15 \cdot 10$ or ________

Area of Triangle

$A = \frac{1}{2}bh$

$A = \frac{1}{2}(5)(4)$ or ________

The area is 150 + 10 or ________ square centimeters.

Check Your Progress Find the area of the figure shown.

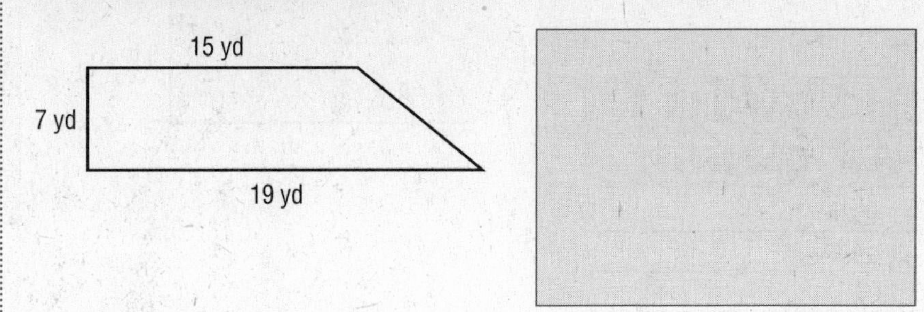

FOLDABLES®

ORGANIZE IT

In the tab for Lesson 11-6 of your Foldable, record in words and symbols how you find the area of composite figures. Make up an example of your own and explain how you would find the area.

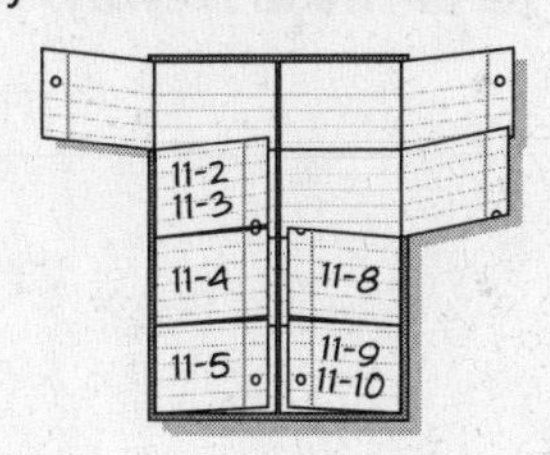

WRITE IT

Explain in general terms how to subdivide a composite figure so you can find its area.

______

______

______

______

EXAMPLE Find the Area of a Composite Figure

2 WINDOWS **The diagram at the right shows the dimensions of a window. Find the area of the window. Round to the nearest tenth.**

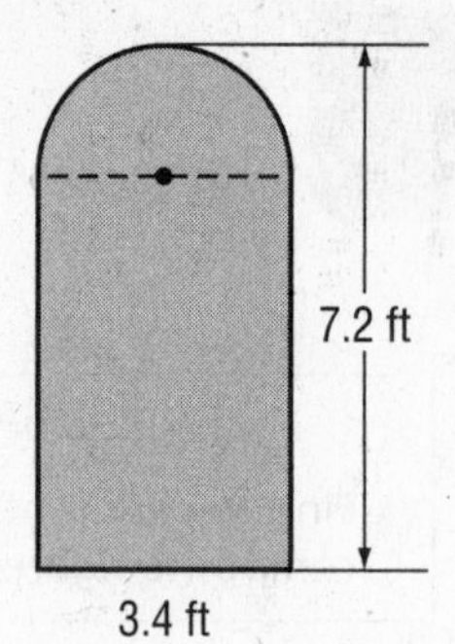

The figure can be separated into a semicircle and a rectangle.

Area of Semicircle

$A = $ ___ πr^2 Area of a semicircle

$A = \frac{1}{2}\pi$ ___   Replace r with ___ ÷ ___ or ___.

$A \approx$ ___   Simplify.

Area of Rectangle

$A = \ell w$ Area of a rectangle

$A = $ ___   Replace ℓ with ___ – ___ or ___

and w with ___.

$A = $ 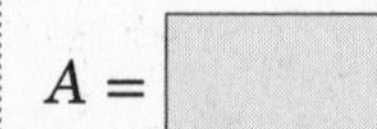   Multiply.

The area of the window is approximately ___ + ___ or ___ square feet.

Check Your Progress The diagram below shows the dimensions of a new driveway. Find the area of the driveway. Round to the nearest tenth.

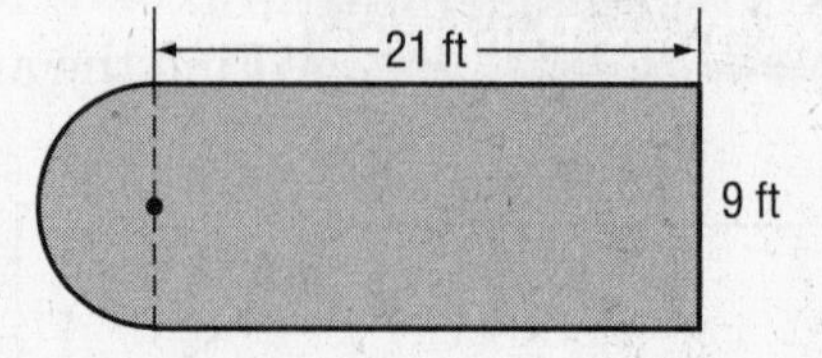

HOMEWORK ASSIGNMENT

Page(s):

Exercises:

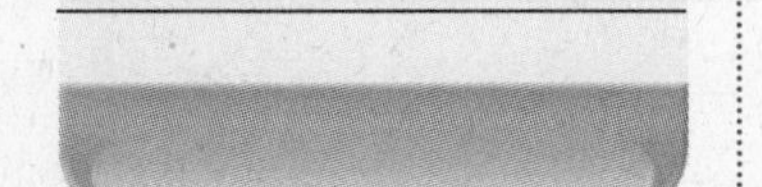

11–7 Three-Dimensional Figures

BUILD YOUR VOCABULARY (pages 255–256)

A **three-dimensional figure** has length, **width**, and depth.

A **face** is a flat ______. The **edges** are the segments formed by intersecting ______. The edges ______ at the **vertices**. The ______ are called **lateral faces.**

MAIN IDEA

- Classify three-dimensional figures.

EXAMPLES Classify Three-Dimensional Figures

For each figure, identify the shape of the base(s). Then classify the figure.

1 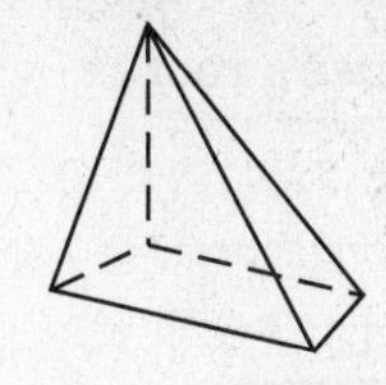

The figure has four triangular faces and one rectangular base. The figure is a ______.

2 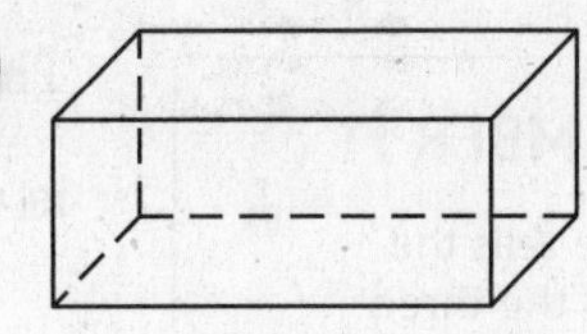

The base and all other faces are rectangles. The figure is a ______.

FOLDABLES®

ORGANIZE IT

Record notes about classifying three-dimensional figures under the tab for Lesson 11-7 of your Foldable.

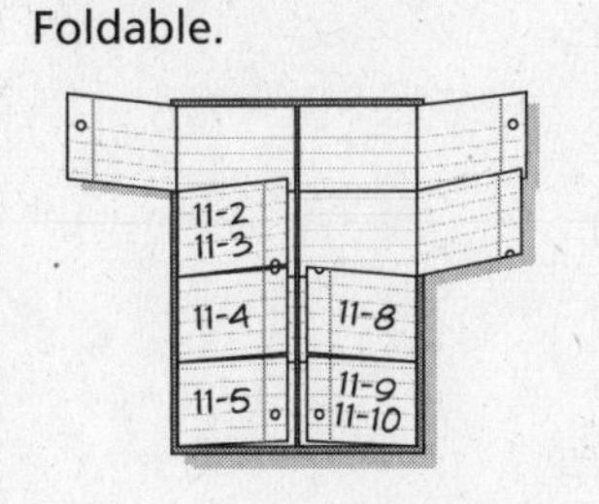

Check Your Progress **For each figure, identify the shape of the base(s). Then classify the figure.**

a. 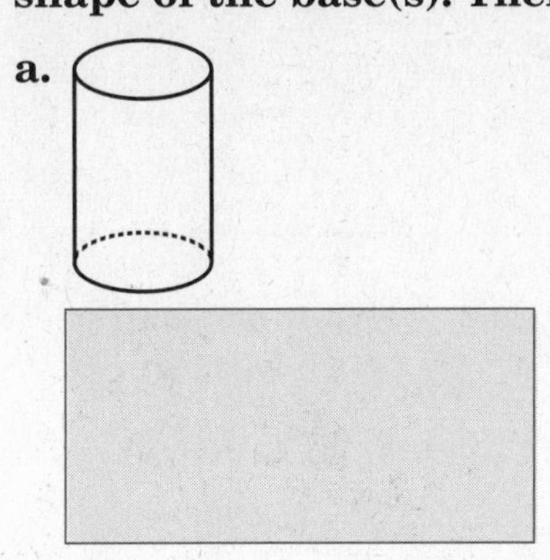

b. 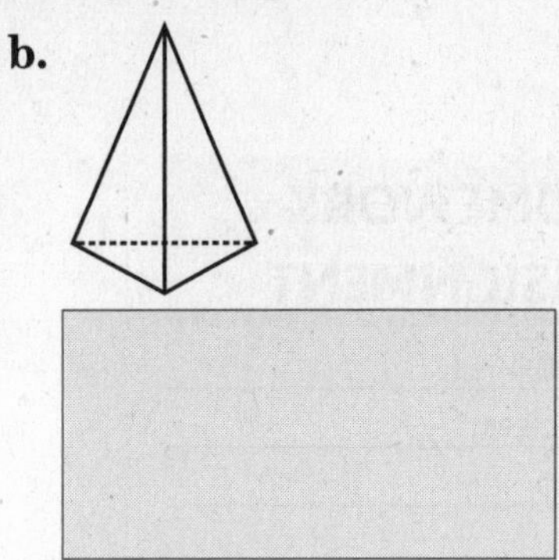

BUILD YOUR VOCABULARY (pages 255–256)

The top and bottom faces of a three-dimensional figure are called the **bases**.

A **prism** has at least three lateral faces that are rectangles.

A **pyramid** has at least three lateral faces that are **triangles**.

A **cone** has one base that is a ________ and one vertex.

A **cylinder** has two bases that are ________ circles.

All of the points on a **sphere** are the same distance from the **center**.

EXAMPLE

3 HOUSES Classify the shape of the house's roof as a three-dimensional figure.

The shape of the house's roof is a ________.

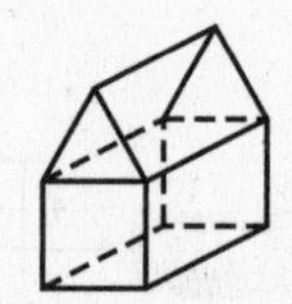

REMEMBER IT

The base tells the name of the three-dimensional figure.

Check Your Progress Classify the shape of the house above, not including the roof.

HOMEWORK ASSIGNMENT

Page(s):

Exercises:

11–8 Drawing Three-Dimensional Figures

MAIN IDEA

- Draw a three-dimensional figure given the top, side, and front views.

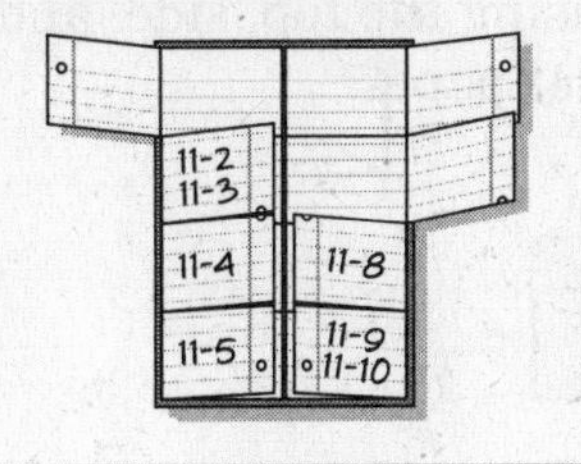

FOLDABLES

ORGANIZE IT

Record notes about drawing three-dimensional figures under the tab for Lesson 11-8 in your Foldable. Sketch examples of rectangular prisms and cylinders.

EXAMPLE Draw a Three-Dimensional Figure

1 Draw a top, a side, and a front view of the figure below.

The top and front views are ________. The side view is a ________.

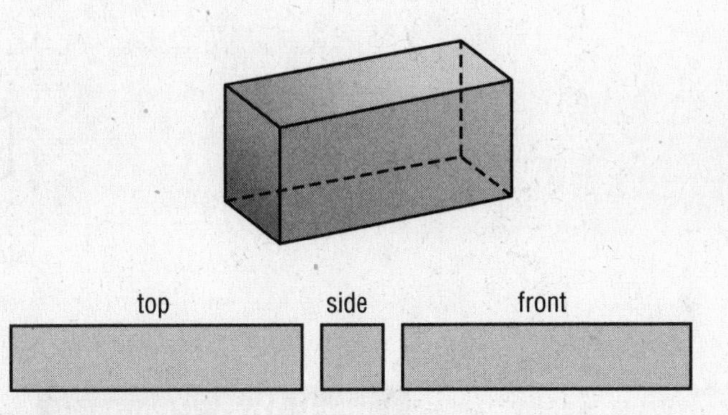

Check Your Progress Draw a top, a side, and a front view of the figure below.

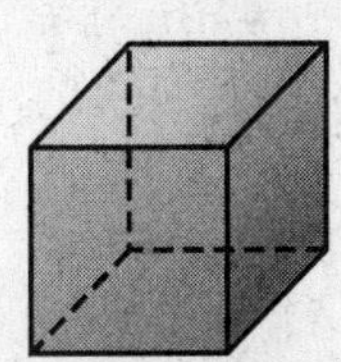

EXAMPLE Draw a Three-Dimensional Figure

2 Draw the three-dimensional figure whose top, side, and front views are shown below. Use isometric dot paper.

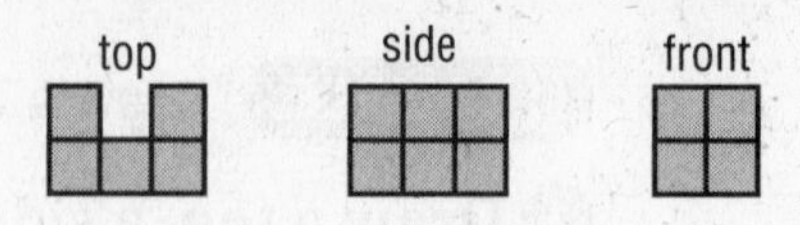

Step 1 Use the top view to draw the base of the figure.

Step 2 Add edges to make the base a solid figure.

Step 3 Use the side and front views to complete the figure.

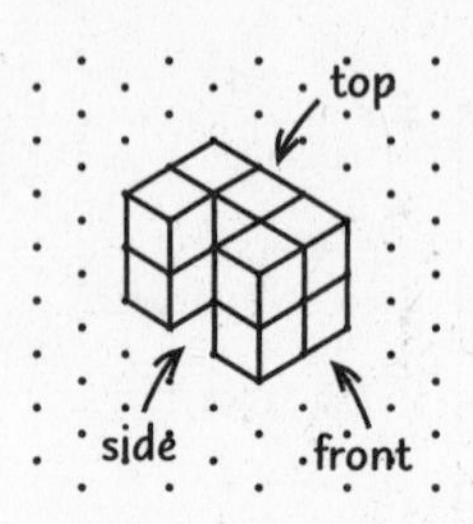

REMEMBER IT

There is more than one way to draw the different views of a three-dimensional figure.

Check Your Progress Draw a solid using the top, side, and front views shown below. Use isometric dot paper.

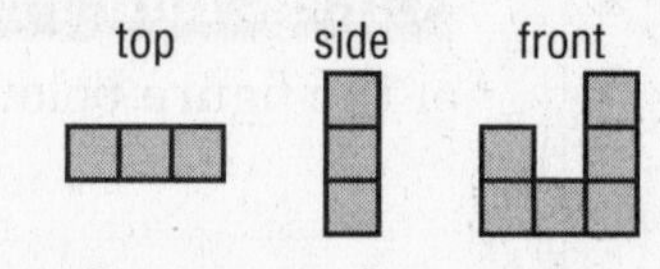

HOMEWORK ASSIGNMENT

Page(s):

Exercises:

11–9 Volume of Prisms

MAIN IDEA

- Find the volumes of rectangular and triangular prisms.

BUILD YOUR VOCABULARY (pages 255–256)

A **volume** of a three-dimensional figure is the measure of ______ occupied by it.

A **rectangular prism** is a prism that has **rectangular** ______. A **triangular prism** has ______ bases.

EXAMPLE Volume of a Rectangular Prism

1 Find the volume of the rectangular prism.

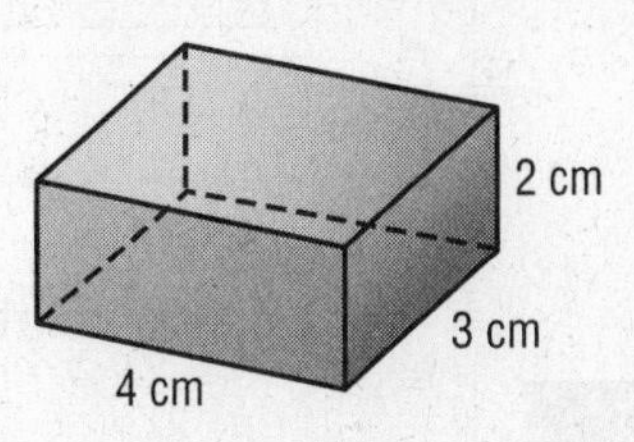

KEY CONCEPT

Volume of a Rectangular Prism The volume V of a rectangular prism is the area of the base B times the height h. It is also the product of the length ℓ, the width w, and the height h.

$V = \ell wh$ Volume of a 

$V =$ ______ Replace ℓ with ______, w with ______, and h with ______.

$V =$ ______ Multiply.

The volume is 24 ______ centimeters.

Check Your Progress Find the volume of the rectangular prism.

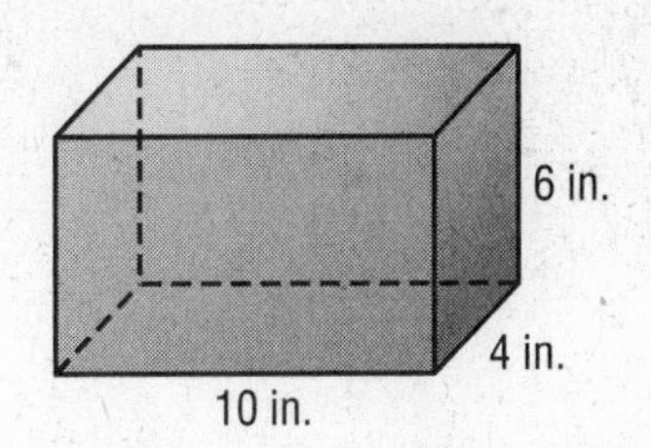

HOMEWORK ASSIGNMENT

Page(s):

Exercises:

11-10 Volume of Cylinders

EXAMPLE Find the Volume of a Cylinder

MAIN IDEA

- Find the volumes of cylinders.

1 Find the volume of the cylinder. Round to the nearest tenth.

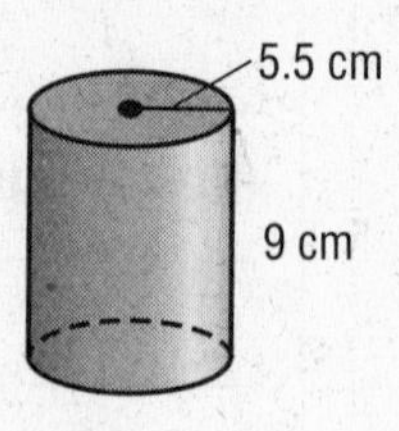

$V =$ ______ Volume of a cylinder

$V = \pi$ ______ Replace the variables.

$V \approx$ ______ Use 3.14 for π.

The volume is about ______ cubic centimeters.

KEY CONCEPT

Volume of a Cylinder The volume V of a cylinder with radius r is the area of the base B times the height h.

FOLDABLES Take notes on how to find the volume of cylinders under the tab for Lesson 11-10 of your Foldable.

Check Your Progress Find the volume of the cylinder. Round to the nearest tenth.

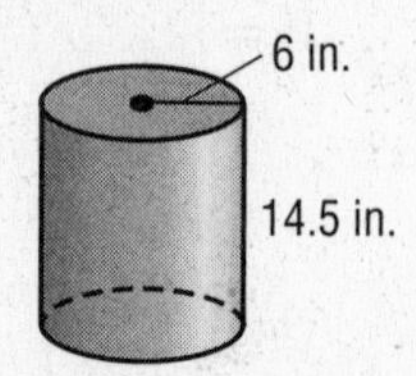

EXAMPLE

2 COFFEE How much coffee can the can hold?

Finest Coffee
6 in.
3 in.

$V = \pi r^2 h$ Volume of a cylinder

$V = \pi(\quad)^2 \quad$ Replace r with ___ and h with ___.

$V \approx$ ___ Simplify.

The coffee can holds about ___ cubic inches.

Check Your Progress **JUICE** Find the volume of a cylinder-shaped juice can that has a diameter of 5 inches and a height of 8 inches.

WRITE IT

Explain how you would use a calculator to evaluate a power.

HOMEWORK ASSIGNMENT

Page(s):

Exercises:

CHAPTER 11

BRINGING IT ALL TOGETHER

STUDY GUIDE

FOLDABLES	VOCABULARY PUZZLEMAKER	BUILD YOUR VOCABULARY
Use your **Chapter 11 Foldable** to help you study for your chapter test.	To make a crossword puzzle, word search, or jumble puzzle of the vocabulary words in Chapter 11, go to: glencoe.com	You can use your completed **Vocabulary Builder** (*pages 255–256*) to help you solve the puzzle.

11-1 Area of Parallelograms

State whether each sentence is *true* or *false*. If false, replace the underlined word to make a true sentence.

1. To find the base of a parallelogram, draw a segment perpendicular to the base with endpoints on opposite sides of the parallelogram. ______

2. The area of a parallelogram is found by multiplying its base times the height. ______

3. What is the area of a parallelogram with a base of 15 feet and a height of 3.5 feet? ______

11-2 Area of Triangles and Trapezoids

Complete the sentence.

4. To find the ______ of a triangle, find the distance from the ______ to the ______ vertex.

Find the area.

5. 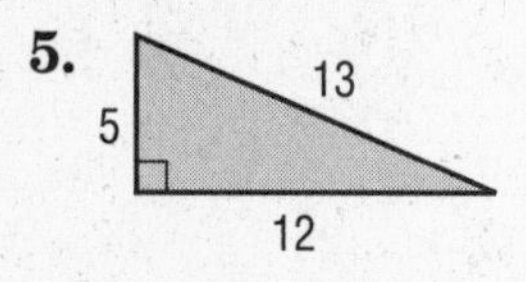

6. 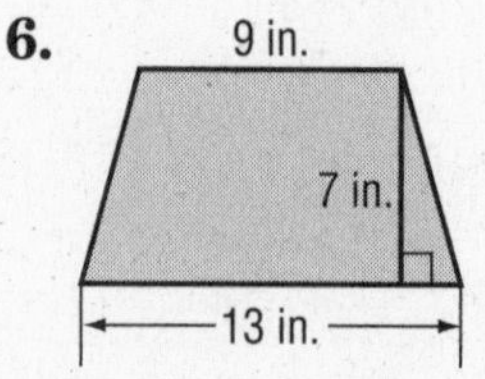

11-3

Circles and Circumference

Find the circumference of each circle. Use 3.14 or $\frac{22}{7}$ for π. Round to the nearest tenth if necessary.

7. radius = 7.4 cm

8. radius = $3\frac{1}{2}$ in.

9. diameter = $6\frac{1}{8}$ ft

10. diameter = 1.7 mi

11-4

Area of Circles

Complete each sentence.

11. To find the ______ of a circle when you are given the ______, divide the length of the diameter by ______, square that, and ______ the result by pi.

12. The units for the ______ of a circle will always be measured in ______ units.

13. Find the area of a circle with a diameter of 13.6 inches. Round to the nearest tenth. ______

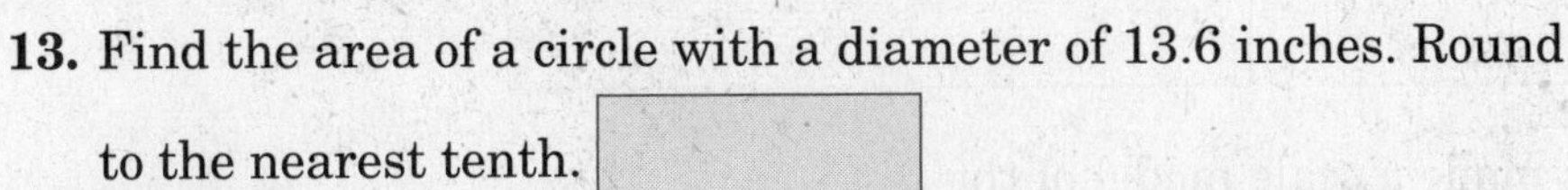

11-5

Problem-Solving Investigation: Solve a Simpler Problem

14. MOVIES Five friends, Marcy, Luke, Shawnda, Jorge, and Lily sat in a row at the movie theater. Marcy and Luke sat next to each other, Jorge did not sit next to Luke, and Shawnda sat at the right end. If Lily sat next to Shawnda and Jorge, find the order of the friends' seating from left to right.

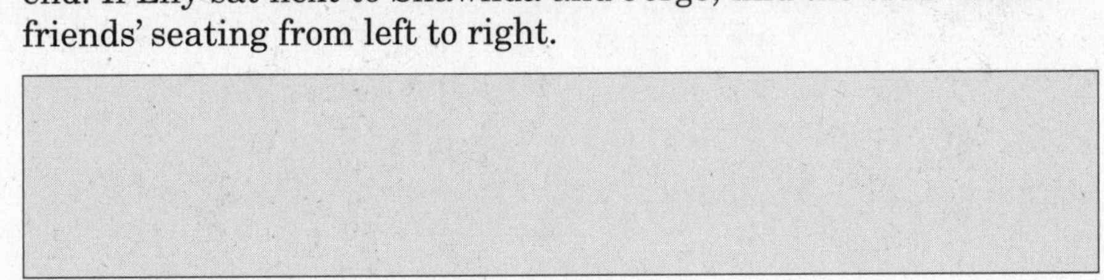

11-6

Area of Composite Figures

Name the two dimensions of the following figures.

15. rectangle

16. triangle

Find the area of each figure. Round to the nearest tenth if necessary.

17.

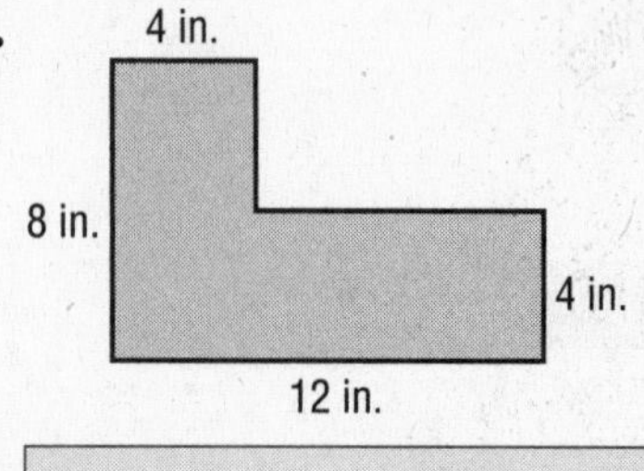

18.

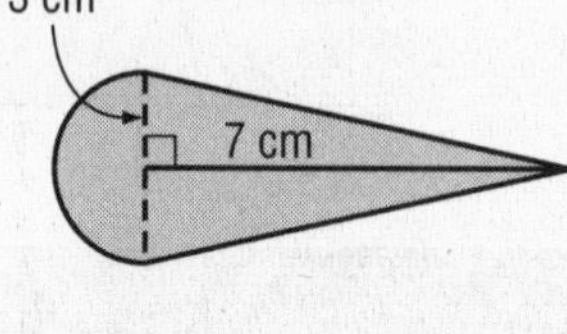

11-7

Three-Dimensional Figures

For each figure, identify the shape of the base(s). Then classify the figure.

19.

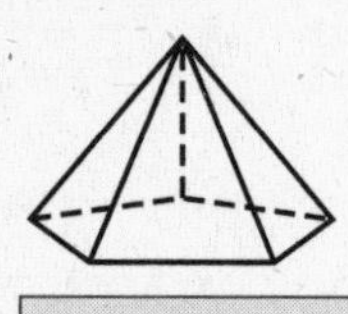

20.

21. MONUMENTS Ginger made a scale model of the Washington Monument as shown. What geometric figure is represented by the top figure of the monument?

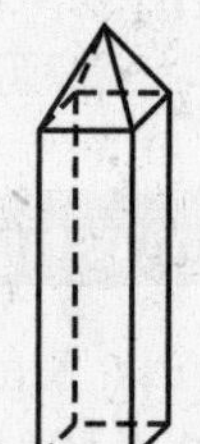

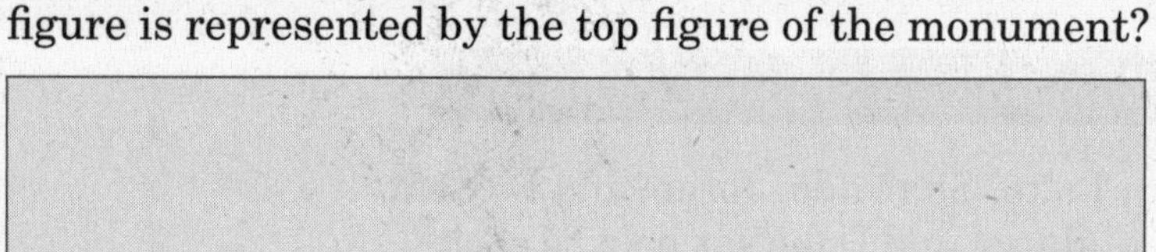

11-8 Drawing Three-Dimensional Figures

Complete each sentence.

22. A two-dimensional figure has two dimensions: ______ and ______.

23. A three-dimensional figure has three dimensions: ______, ______ and ______.

11-9 Volume of Prisms

Find the volume of rectangular prisms with these dimensions. Round to the nearest tenth if necessary.

24. 4 ft by 12 ft by 7 ft

25. 9 in. by 8 in. by 5.5 in.

26. 2.5 in. by 6 in. by 5 in.

27. 3.8 cm by 2.4 cm by 2 cm

11-10 Volume of Cylinders

Write C if the phrase is true of a cylinder, P if it is true of a prism, and CP if the phrase is true of both.

28. ______ has bases that are parallel and congruent.

29. ______ has sides and bases that are polygons.

30. ______ has bases that are circular.

31. ______ is a solid.

32. ______ has volume.

33. ______ is three-dimensional.

ARE YOU READY FOR THE CHAPTER TEST?

Math Online

Visit **glencoe.com** to access your textbook, more examples, self-check quizzes, and practice tests to help you study the concepts in Chapter 11.

Check the one that applies. Suggestions to help you study are given with each item.

☐ **I completed the review of all or most lessons without using my notes or asking for help.**

- You are probably ready for the Chapter Test.
- You may want to take the Chapter 11 Practice Test on page 631 of your textbook as a final check.

☐ **I used my Foldable or Study Notebook to complete the review of all or most lessons.**

- You should complete the Chapter 11 Study Guide and Review on pages 626–630 of your textbook.
- If you are unsure of any concepts or skills, refer back to the specific lesson(s).
- You may want to take the Chapter 11 Practice Test on page 631 of your textbook.

☐ **I asked for help from someone else to complete the review of all or most lessons.**

- You should review the examples and concepts in your Study Notebook and Chapter 11 Foldable.
- Then complete the Chapter 11 Study Guide and Review on pages 626–630 of your textbook.
- If you are unsure of any concepts or skills, refer back to the specific lesson(s).
- You may also want to take the Chapter 11 Practice Test on page 631 of your textbook.

Student Signature

Parent/Guardian Signature

Teacher Signature

CHAPTER 12

Geometry and Measurement

Use the instructions below to make a Foldable to help you organize your notes as you study the chapter. You will see Foldable reminders in the margin of this Interactive Study Notebook to help you in taking notes.

Begin with a sheet of 11" by 17" paper.

STEP 1 **Fold** the paper in fourths lengthwise.

STEP 2 **Fold** a 2" tab along the short side. Then fold the rest in half.

STEP 3 **Draw** lines along folds and label as shown.

Ch. 12	Rectangular Prisms	Cylinders
Draw Examples		
Find Volume		
Find Surface Area		

NOTE-TAKING TIP: When taking notes about 3-dimensional figures, it is important to draw examples. It also helps to record any measurement formulas.

BUILD YOUR VOCABULARY

This is an alphabetical list of new vocabulary terms you will learn in Chapter 12. As you complete the study notes for the chapter, you will see Build Your Vocabulary reminders to complete each term's definition or description on these pages. Remember to add the textbook page number in the second column for reference when you study.

Vocabulary Term	Found on Page	Definition	Description or Example
hypotenuse			
irrational number			
leg			
Pythagorean Theorem			
surface area			

12-1 Estimating Square Roots

MAIN IDEA

- Estimate square roots.

EXAMPLE Estimate the Square Root

1 **Estimate $\sqrt{96}$ to the nearest whole number.**

List some perfect squares.

1, 4, 9, 16, 25, 36, 49, 64, 81, 100,...

$81 < 96 < 100$ 96 is between the ______

squares ______ and ______.

$\_\_\_ < \sqrt{96} < \_\_\_$ Find the $\sqrt{}$ of each number.

$\_\_\_ < \sqrt{96} < \_\_\_$ ______ = 9 and

______ = 10

So, $\sqrt{96}$ is between ______ and ______. Since 96 is closer to ______ than 81, the best whole number estimate is ______. Verify with a calculator.

Check Your Progress **Estimate each square root to the nearest whole number.**

a. $\sqrt{41}$

b. $\sqrt{86}$

c. $\sqrt{138}$

BUILD YOUR VOCABULARY (page 280)

A number that cannot be written as a  is an **irrational number**.

EXAMPLE Use a Calculator to Estimate

REMEMBER IT

Decimals used to represent irrational numbers are estimates, not exact values.

2 **Graph $\sqrt{37}$ on a number line.**

2nd [$\sqrt{\ }$] 37 ENTER 

$\sqrt{37} \approx$ ____

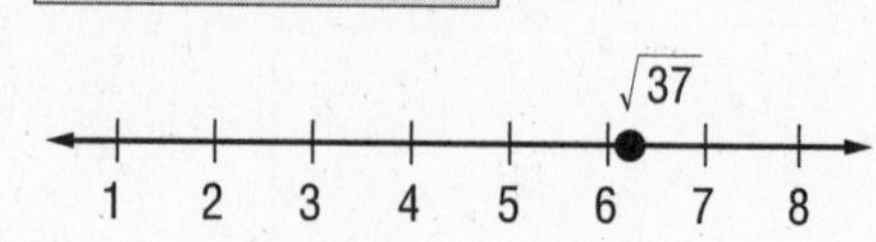

Check ____ = 36 and ____ = 49. Since ____ is between 36 and 49, the answer, ____, is reasonable.

Check Your Progress **Graph each number on a number line.**

a. $\sqrt{78}$

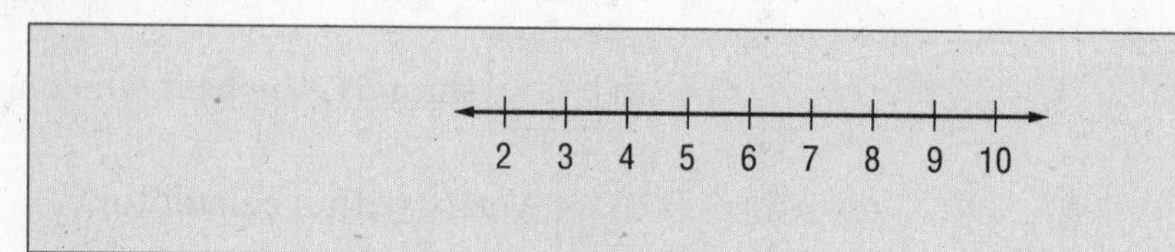

b. $\sqrt{96}$

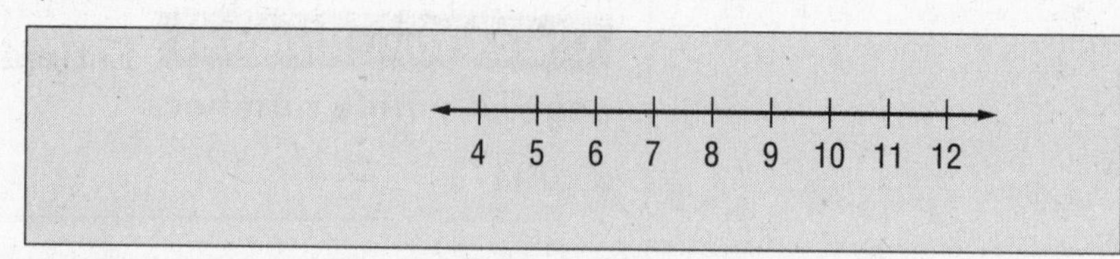

c. $\sqrt{188}$

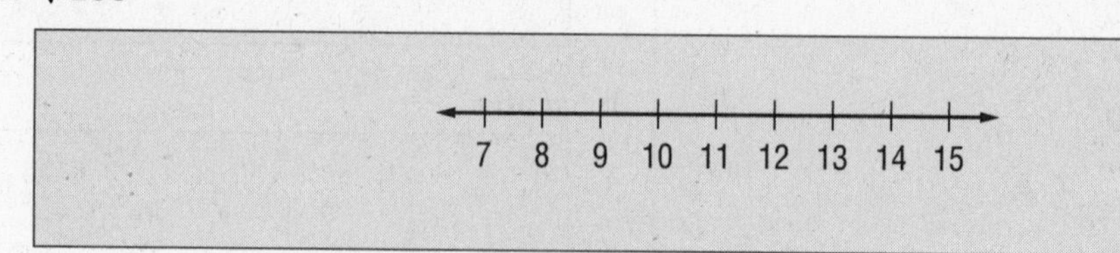

HOMEWORK ASSIGNMENT

Page(s):

Exercises:

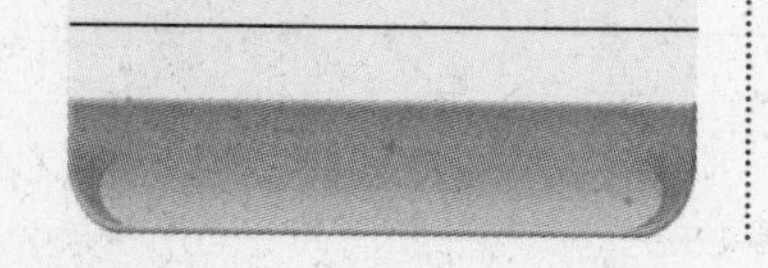

12-2 The Pythagorean Theorem

MAIN IDEA

- Find length using the Pythagorean Theorem.

BUILD YOUR VOCABULARY (page 280)

The two sides ________ to the right ________ of a right triangle are the **legs**.

The side ________ the right ________ of a right triangle is the **hypotenuse**.

The **Pythagorean Theorem** describes the relationship between the length of the ________ and the lengths of the ________.

KEY CONCEPT

Pythagorean Theorem In a right triangle, the square of the length of the hypotenuse equals the sum of the squares of the lengths of the legs.

EXAMPLE Find the Length of the Hypotenuse

1 Find the length of the hypotenuse of the triangle.

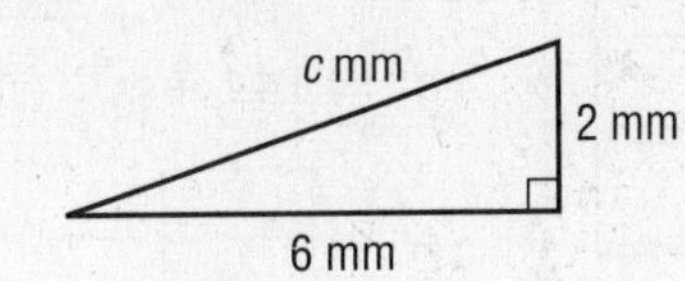

$c^2 = a^2 + b^2$	Pythagorean Theorem
$c^2 =$ ________	Replace *a* with 2 and *b* with 6.
$c^2 = 4 + 36$	Evaluate 2^2 and 6^2.
$c^2 = 40$	Add.
$c = \pm\sqrt{40}$	Definition of square root
$c = \pm 6.3$	Simplify.

The length of the hypotenuse is about ________ millimeters.

Check Your Progress Find the length of the hypotenuse of a right triangle if the legs are 5 centimeters and 7 centimeters.

REVIEW IT

How do you know if a triangle is a right triangle? *(Lesson 10-4)*

EXAMPLE

2 SPORTS A gymnastics tumbling floor is in the shape of a square. If a gymnast flips from one corner to the opposite corner, about how far has he flipped?

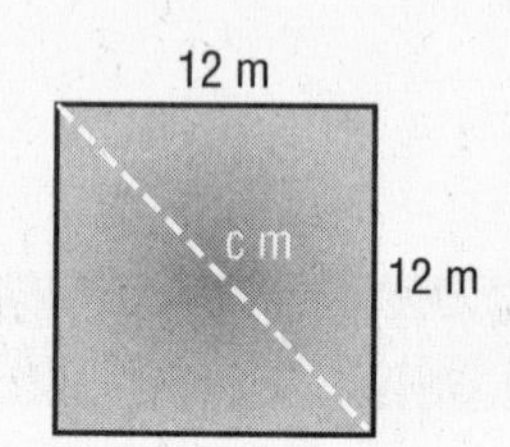

To solve, find the length of the hypotenuse c.

$c^2 = a^2 + b^2$	Pythagorean Theorem
$c^2 =$ ☐ $+ 12^2$	Replace a with ☐ and b with ☐.
$c^2 = 144 +$ ☐	Evaluate ☐.
$c^2 =$ ☐	Add.
$\sqrt{c^2} = \pm$ ☐	Take the  of each side.
$c \approx \pm$ ☐	Simplify.

The gymnast will have flipped about ☐.

Check Your Progress **SEWING** Rose has a rectangular piece of fabric 28 inches long and 16 wide. She wants to decorate the fabric with lace sewn across both diagonals. How much lace will Rose need?

EXAMPLE Find the Length of a Leg

3 Find the missing measure of the triangle at the right.

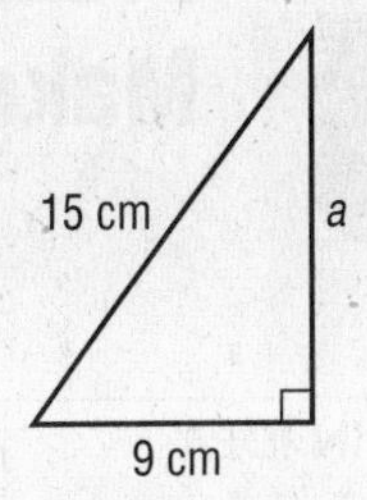

$c^2 = a^2 + b^2$	Pythagorean Theorem
$\boxed{}^2 = a^2 + \boxed{}^2$	Replace b with ___ and c with ___.
$\boxed{} = a^2 + \boxed{}$	Evaluate ___ and ___.
$225 - \boxed{} = a^2 + 81 - \boxed{}$	Subtract ___ from each side.
$\boxed{} = a^2$	Simplify.
$\sqrt{144} = \sqrt{a^2}$	Take the ___ of each side.
$\boxed{} = a$	Simplify.

The length of the leg is ___ centimeters.

Check Your Progress Find the missing measure of the triangle. Round to the nearest tenth if necessary.

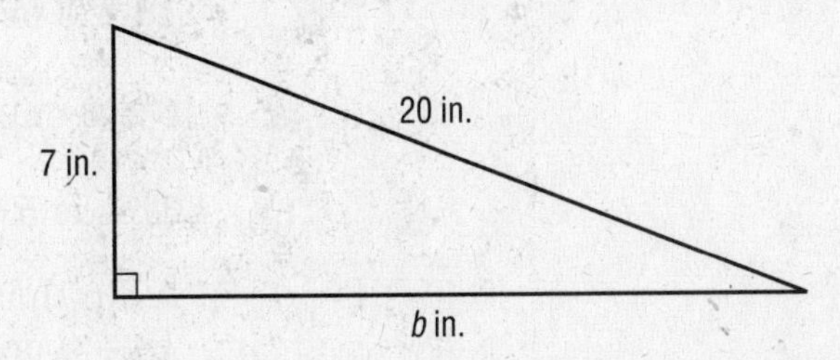

HOMEWORK ASSIGNMENT

Page(s):

Exercises:

12–3 Problem-Solving Investigation: Make a Model

MAIN IDEA

- Solve problems by making a model.

EXAMPLE Make a Model to Solve the Problem

STORAGE **A daycare center plans to make simple wooden storage bins for the 3-inch square alphabet blocks. If each bin will hold 30 blocks, give two possible dimensions for the inside of the bin.**

UNDERSTAND You know the dimensions of the blocks and that each bin holds 30 blocks. You need to give two possible dimensions for the inside of the bin.

PLAN Make a cardboard model of a cube with sides 3 inches long. Then use your model to determine the dimensions of the bin that will hold 30 cubes.

SOLVE

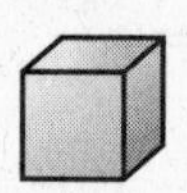

A bin that holds 5 cubes in length, 3 cubes in width, and 2 cubes in height would hold 30 cubes. This bin would be 15 inches in length, 9 inches in width, and 6 inches in height. A bin that holds 6 cubes in length, 5 cubes in width, and 1 cube in height would also hold 30 cubes. This bin would be 18 inches in length, 15 inches in width, and 3 inches in height.

CHECK A bin that is 15 in. × 9 in. × 6 in. would hold $15 \div 3$ or ☐ cubes by $9 \div 3$ or 3 cubes by $6 \div 3$ or ☐ cubes in height.

This is $5 \times 3 \times 2$ or ☐ cubes.

A bin that is 18 in. × 15 in. × 3 in. would hold $18 \div 3$ or 6 cubes by $15 \div 3$ or 5 cubes by $3 \div 3$ or 1 cube. This is $6 \times 5 \times 1$ or 30 cubes.

Check Your Progress **FRAMES** A photo that is 5 inches by 7 inches will be placed in a frame that has a metal border of 1.5 inches on each side. What are the dimensions of the frame?

HOMEWORK ASSIGNMENT

Page(s):

Exercises:

12–4 Surface Area of Rectangular Prisms

BUILD YOUR VOCABULARY (page 280)

The ______ of the areas of all of the ______, or faces, of a ______ figure is the **surface area.**

MAIN IDEA

- Find the surface areas of rectangular prisms.

KEY CONCEPT

Surface Area of Rectangular Prisms
The surface area S of a rectangular prism with length ℓ, width w, and height h is the sum of the areas of the faces.

EXAMPLE Use a Net to Find Surface Area

1 Find the surface area of the rectangular prism.

You can use a net of the rectangular prism to find its surface area. There are three pairs of congruent faces.

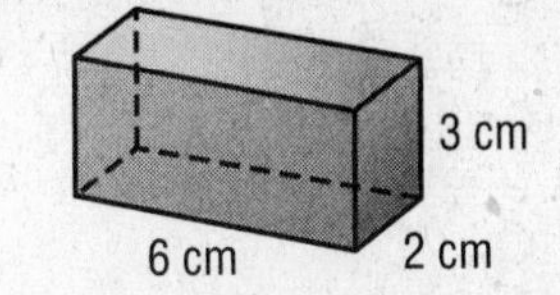

- top and bottom
- front and back
- two sides

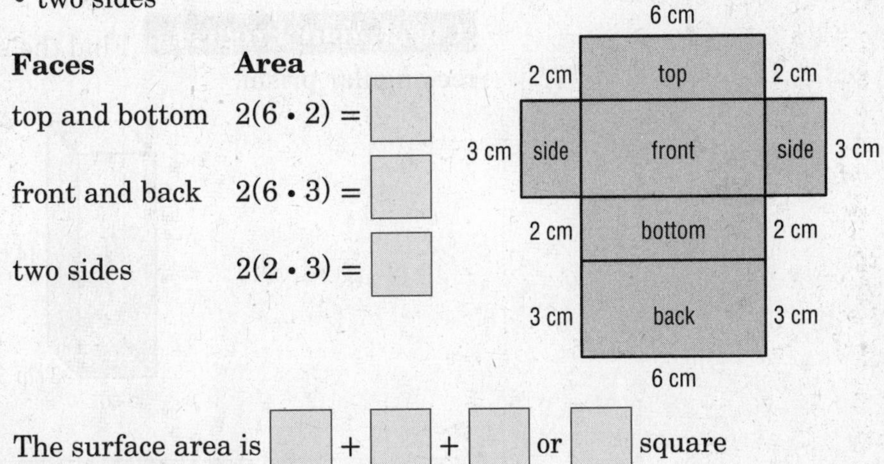

Faces	Area
top and bottom	$2(6 \cdot 2) =$ ______
front and back	$2(6 \cdot 3) =$ ______
two sides	$2(2 \cdot 3) =$ ______

The surface area is ______ + ______ + ______ or ______ square centimeters.

Check Your Progress Find the surface area of the rectangular prism.

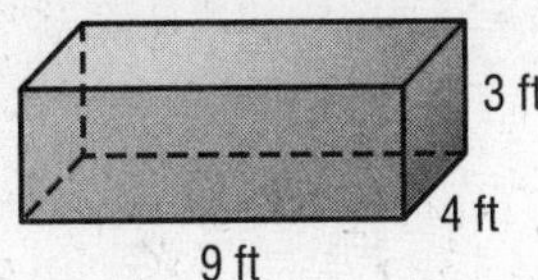

FOLDABLES

ORGANIZE IT

Include information in words and symbols on how to find the surface area of rectangular prisms in the appropriate section of your Foldable table.

Ch. 12	Rectangular Prisms	Cylinders
Draw Examples		
Find Volume		
Find Surface Area		

EXAMPLE **Use a Formula to Find Surface Area**

2 Find the surface area of the rectangular prism.

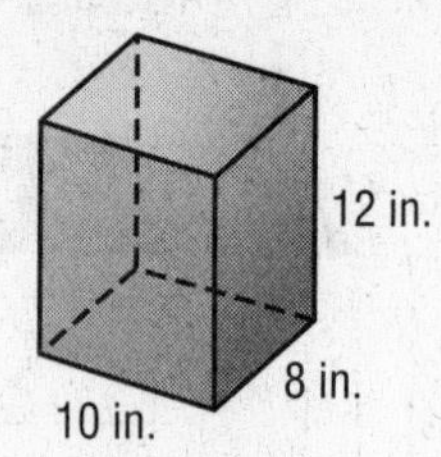

Replace ℓ with ___, w with ___, and h with ___.

surface area $= 2\ell w + 2\ell h + 2wh$

$=$ ___ $+$ ___ $+$ ___

$=$ ___ $+$ ___ $+$ ___ Multiply first. Then add.

$=$ ___

The surface area of the prism is ___.

Check Your Progress Find the surface area of the rectangular prism.

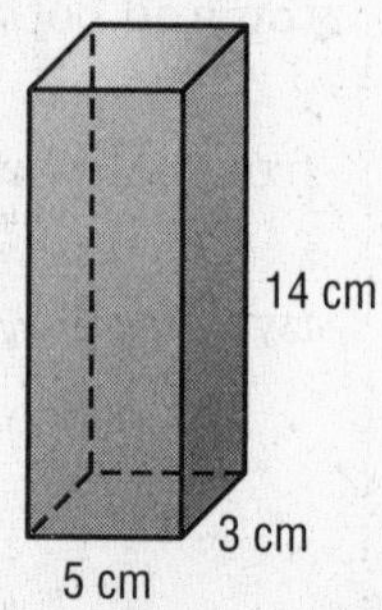

EXAMPLE

3 BOXES Drew is putting together a cardboard box that is 9 inches long, 6 inches wide, and 8 inches high. He bought a roll of wrapping paper that is 1 foot wide and 3 feet long. Did he buy enough to wrap the box? Explain.

Step 1 Find the surface area of the box.

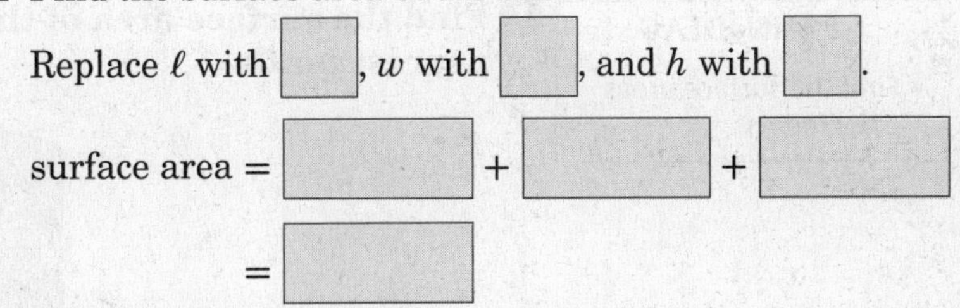

Replace ℓ with ______, w with ______, and h with ______.

surface area = ______ + ______ + ______

= ______

Step 2 Find the area of the wrapping paper.

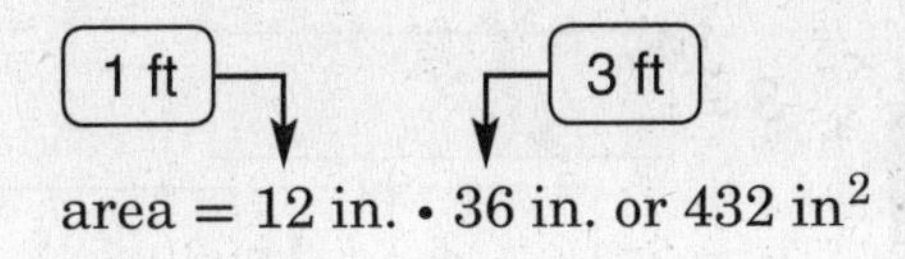

area = 12 in. • 36 in. or 432 in^2

Since 432 ______ 348, Drew bought enough wrapping paper.

Check Your Progress **FABRIC** Angela needs to cover a cardboard box that is 15 inches long, 5 inches wide, and 4 inches high with felt. She bought a piece of felt that is 1 foot wide and $2\frac{1}{2}$ feet long. Did she buy enough felt to cover the box? Explain.

HOMEWORK ASSIGNMENT

Page(s):

Exercises:

12-5 Surface Area of Cylinders

MAIN IDEA

- Find the surface areas of cylinders.

KEY CONCEPT

Surface Area of a Cylinder The surface area S of a cylinder with height h and radius r is the sum of the areas of circular bases and the area of the curved surface.

EXAMPLE Find Surface Area of a Cylinder

1 **Find the surface area of the cylinder. Round to the nearest tenth.**

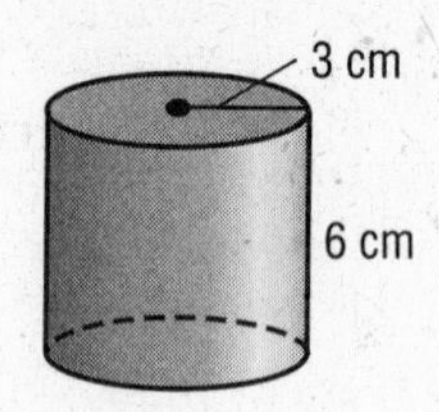

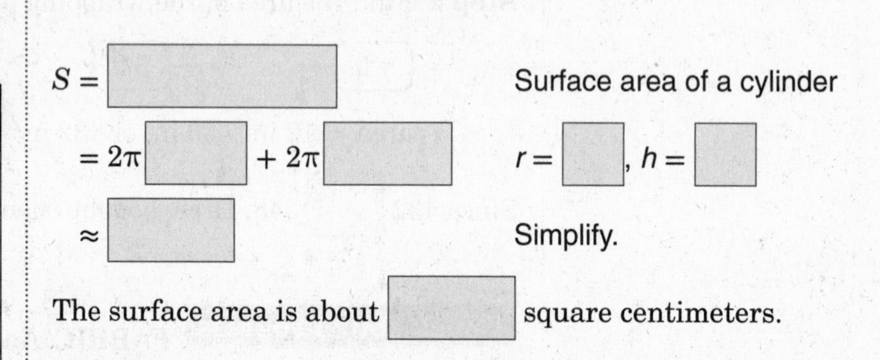

$S =$ ______ Surface area of a cylinder

$= 2\pi$ ______ $+ 2\pi$ ______ $r =$ ______, $h =$ ______

$\approx$ ______ Simplify.

The surface area is about ______ square centimeters.

EXAMPLE

2 GIFT WRAP **A poster is contained in a cardboard cylinder that is 10 inches high. The cylinder's base has a diameter of 8 inches. How much paper is needed to wrap the cardboard cylinder if the ends are to be left uncovered?**

Since only the curved side of the cylinder is to be covered, you do not need to include the areas of the top and bottom of the cylinder.

$S =$ ______ Curved surface of a cylinder

$=$ ______ $r = 4$, $h = 10$

$\approx$ ______ Simplify.

About 251.3 ______ of paper is needed.

FOLDABLES

ORGANIZE IT

Include information in words and symbols about how to find the surface area of a cylinder in the appropriate section of your Foldable table.

Ch. 12	Rectangular Prisms	Cylinders
Draw Examples		
Find Volume		
Find Surface Area		

HOMEWORK ASSIGNMENT

Page(s):

Exercises:

Check Your Progress

a. Find the surface area of the cylinder. Round to the nearest tenth.

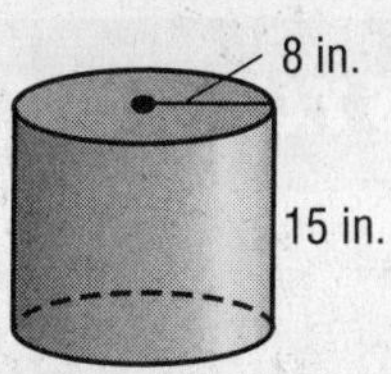

b. LABELS A can of fruit juice is in the shape of a cylinder with a diameter of 6 inches and a height of 12 inches. How much paper is needed to create the label if the ends are to be left uncovered?

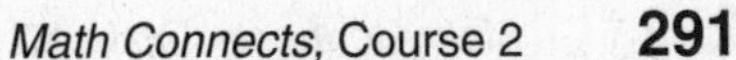

CHAPTER 12

BRINGING IT ALL TOGETHER

STUDY GUIDE

FOLDABLES	VOCABULARY PUZZLEMAKER	BUILD YOUR VOCABULARY
Use your **Chapter 12 Foldable** to help you study for your chapter test.	To make a crossword puzzle, word search, or jumble puzzle of the vocabulary words in Chapter 12, go to: glencoe.com	You can use your completed **Vocabulary Builder** (*page 280*) to help you solve the puzzle.

12-1 Estimating Square Roots

Estimate each square root to the nearest whole number.

1. $\sqrt{95}$

2. $\sqrt{51}$

3. $\sqrt{150}$

4. $\sqrt{230}$

12-2 The Pythagorean Theorem

State whether each sentence is *true* or *false*. If *false*, replace the underlined word to make a true sentence.

5. The Pythagorean Theorem states that $c^2 = a^2 + b^2$, where $\underline{a}$ represents the length of the hypotenuse.

6. The <u>hypotenuse</u> is always the longest of the three sides of a right triangle.

Find the missing measure of each right triangle. Round to the nearest tenth if necessary.

7.

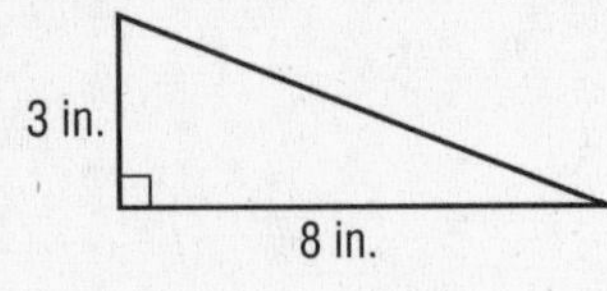

8.

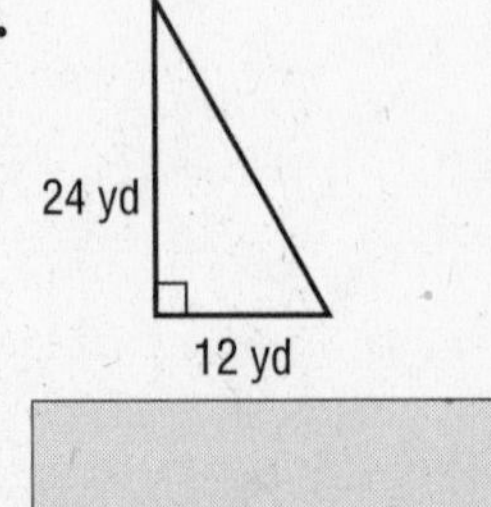

12-3 Problem-Solving Investigation: Make a Model

9. **BOOKS** A bookstore will arrange 4 books in a row in the store window. In how many different ways can the store arrange these 4 books?

12-4 Surface Area of Rectangular Prisms

Find the surface area of each rectangular prism. Round to the nearest tenth if necessary.

10.

3 in.

4 in.

15 in.

11.

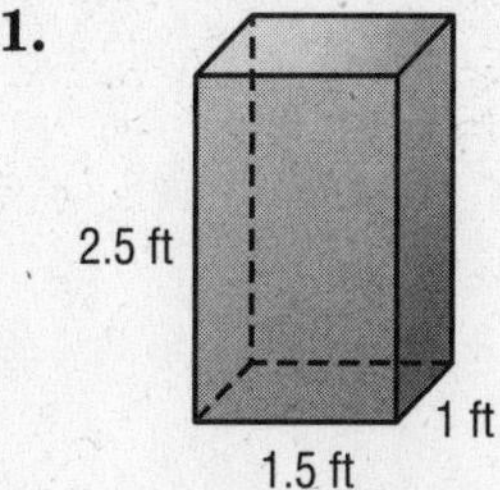

12.

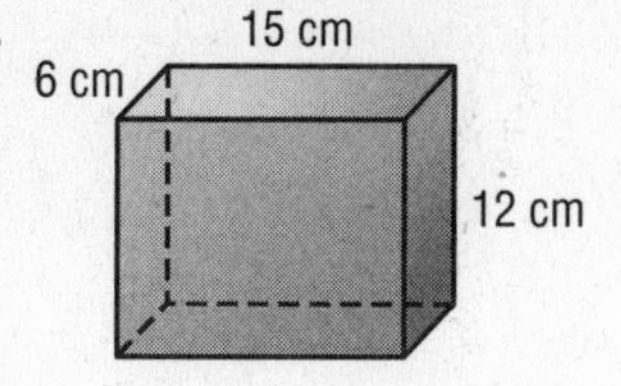

12-5 Surface Area of Cylinders

Write the formula to find each of the following.

13. the area of a circle

14. the circumference of a circle

15. the area of a rectangle

Find the surface area of the cylinder. Round to the nearest tenth if necessary.

16.

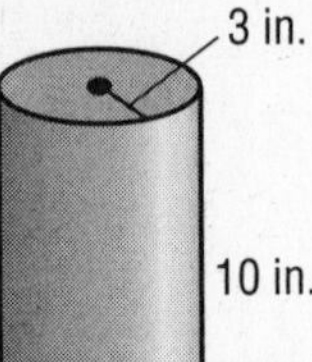

ARE YOU READY FOR THE CHAPTER TEST?

Math Online

Visit **glencoe.com** to access your textbook, more examples, self-check quizzes, and practice tests to help you study the concepts in Chapter 12.

Check the one that applies. Suggestions to help you study are given with each item.

☐ **I completed the review of all or most lessons without using my notes or asking for help.**

- You are probably ready for the Chapter Test.
- You may want to take the Chapter 12 Practice Test on page 663 of your textbook as a final check.

☐ **I used my Foldable or Study Notebook to complete the review of all or most lessons.**

- You should complete the Chapter 12 Study Guide and Review on pages 660–662 of your textbook.
- If you are unsure of any concepts or skills, refer back to the specific lesson(s).
- You may want to take the Chapter 12 Practice Test on page 663 of your textbook.

☐ **I asked for help from someone else to complete the review of all or most lessons.**

- You should review the examples and concepts in your Study Notebook and Chapter 12 Foldable.
- Then complete the Chapter 12 Study Guide and Review on pages 660–662 of your textbook.
- If you are unsure of any concepts or skills, refer back to the specific lesson(s).
- You may also want to take the Chapter 12 Practice Test on page 663 of your textbook.

Student Signature

Parent/Guardian Signature

Teacher Signature

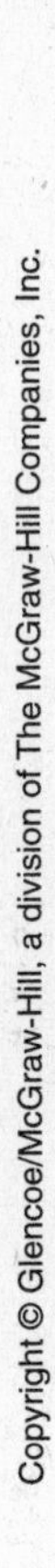